U0324069

帝都来信

北京皇家园林概览

A PARTICULAR ACCOUNT OF THE EMPEROR OF CHINA'S GARDENS NEAR PEKIN

王致诚　著

JEAN DENIS ATTIRET

段建强　译

TRANSLATED BY DUAN JIANQIANG

同济大学出版社

TONGJI UNIVERSITY PRESS

图书在版编目（CIP）数据

帝都来信：北京皇家园林概览 / 王致诚著；段建强译. -- 上海：同济大学出版社，2018.12

ISBN 978-7-5608-8268-0

Ⅰ.①帝… Ⅱ.①王… ②段… Ⅲ.①园林艺术—文化交流—中国、西方国家②绘画技法—文化交流—中国、西方国家 Ⅳ.① TU986.1 ② J21

中国版本图书馆 CIP 数据核字 (2018) 第 289236 号

帝都来信：北京皇家园林概览

王致诚 **著** 段建强 **译**

顾 问：卢永毅
责任编辑：吕 炜
责任校对：徐春莲
装帧设计：完 颖

出版发行：同济大学出版社 www.tongjipress.com.cn
（地址：上海市四平路 1239 号 邮编：200092 电话：021-65985622）
经 销：全国各地新华书店、建筑书店、网络书店
印 刷：上海安枫印务有限公司
开 本：889mm×1 194mm 1/32
印 张：4.75
字 数：128 000
版 次：2018 年 12 月第 1 版 2018 年 12 月第 1 次印刷
书 号：ISBN 978-7-5608-8268-0
定 价：49.00 元

王致诚 Jean Denis Attiret
(1702—1768)

天主教耶稣会传教士，法国人，自幼学画于里昂，后留学罗马。

清乾隆三年(1738)来中国，献《三王来朝耶稣图》，乾隆时受召供奉内廷。初绘西画，然不为清帝所欣赏，后学中国绘画技法，参酌中西画法，别立中西折中之新体，曲尽帝意，乃得重视。与郎世宁、艾启蒙、安德义合称四洋画家，形成新体画风。

序

卢永毅

多年前，我的好友、米兰工学院的青年教师Giancarlo Floridi送给我一件特殊礼物——法国耶稣会传教士王致诚写的*A Particular Account of the Emperor of China's Gardens near Pekin*。这原本是部法语著作，是作者于1743年从他工作的帝都北京写给其巴黎友人的一封长信，于1749年在巴黎出版。Giancarlo了解我的外语能力有限，又很想寻找中西文化交流史上的代表性读物送我，所以选了该书的英译本。这虽然是一个总共只有50页的小册子，但很特别，因为它是由一个叫档案重印公司（Archival Reprint Company）的机构印制，书中夹着的小条注明，这是将1752年伦敦首发的珂罗版英译版扫描后复制而成的，扉页上还有它是当年85册中第21个拷贝的字样。虽然，这个拷贝的原件现藏于哪家图书馆不得而知，但一本泛黄的仿真古籍捧在手中，已经万分欢喜，穿越时空的神奇感受油然而生。惭愧的是，由于自身的惰性和与园林研究的距离，这件可爱的礼物珍藏于我的书柜数年，却实在是形式大于内容。平日偶尔翻上一页就会想，它是在等着更合适的人去细细品它呢！

现在，看到这珍贵的文献已由建强翻译、研究并即将在同济大学出版社出版，真是非常高兴。建强就是这更合适的人，他语文功底好，又是园林研究专家，他将书名译成《帝都来信》，有一种别样的贴切，更关键的是，他完成的译文如此优美流畅，读起来充满了画面感，仿佛时间回流，让我跟随这位乾隆皇帝的宫廷洋画师细细“游览”了一次那座曾经无与伦比的皇家园林——圆明园。当然，文本的字里行间更触动我的，是对这样一位“由万里航海而来”的法国传教士的各种想象，以及有关中西文化交流史研究的更多思考。

基督宗教在近代之前至少有三次入华高潮，在各个方面深刻影响了东西方世界的互相认知和理解，其中最为重要也最为深入的阶段是第三次，始自明末清初直至19世纪前。西方耶稣会士传教士们无疑是这过程中最重要的贡献者，为历史留下了罗明坚、利玛窦、汤若望、金尼阁、张诚、李明、南怀仁、郎世宁、王致诚、蒋友仁等一连串影响深远的人物的名字。在这些人中，如果我们专注于追溯中西建筑与园林文化的交流史，那么以这《帝都来信》为证，王致诚当是其中最凸显的一位。

会有同行说，最早17世纪末，英国人威廉·坦普尔爵士（Sir William Temple）就已经在王致诚之前比较了中国园林与欧洲园林的差异，并表达出对中国园林非规则特征的好感，甚至还为这种“无序美”找了一个“Sharawadgi”的专用词来形象论述。论及与王致诚的同时代人，大家更是认定钱伯斯爵士（Sir William Chambers）的历史地位，称他是真正开启了西方人研究中国建筑与园林的第一人。但现在，当我们有机会流畅地阅读王致诚的这封“长信”时，对历史的触觉还是会被刷新一次。事实上，坦普尔从未到过中国，所以他对中国园林特征的“发现”得归功于另一位来自那不勒斯的耶稣会传教士马国贤，马国贤受康熙旨意完成的铜版画《御制避暑山庄图咏三十六景》（1710s），首次以中西合璧的风格将中国的皇家园林展示给了欧洲人。至于钱伯斯，事实上当他动身来中国时，王致诚已在为乾隆皇帝的圆明园建造西洋楼，并在忙碌之余完成了这封长信，而且，与王致诚能在帝都享有亲眼目睹皇家园林的特权相比，钱伯斯是奢望不及的。

其实，这里无意要用如此简单化的叙述来比孰高孰低，而是想说，东、西方文明的相遇，有很多历史的必然与巧合，但无论历史如何宏阔地展开，离开具体而微的历史人物卓绝的努力，这种相遇都难免会失去应有的丰富性，甚至还会陷入偏见而浑然不觉。回到王致诚的这封书信，应该说很难再有比这更接近历史的一手资料了。看书信的丰富记述，当时的这位传教士显然是被这“在世界其他地方都未曾见过”的园林胜景深深吸引，身为画家，他何以不想把亲眼所见逐一绘制。但正如他坦言，对于这景致如此丰富的皇宫御园，“即便我终年无事，专事此图，也至少要画上三年的时日”。于是，王致诚转而利用一切闲暇，以书信形式将所见所闻一一描绘。而在我看来，他的文字叙述恰能在当时向西方人介绍中国建筑与园林特征的过程中显现出一种独特的魅力。王致诚着实领会到，这看似随意布局的园林实际是被精心营造的，因而这需要在时间的延展和观者的游走中才能“渐次呈现”的丰富景致，事实上只有“身临其境”才是“唯一能感受它的方式”。那么，面对种种条件的限制，除了尽可能地用这细腻真切的文字描述，还有更好的方式去向遥远的欧洲人传达身在其中的精妙体验吗？反过来，现在有这些入微的叙述，尤其像对园内取悦帝后嫔妃的市集活动的描写，活灵活现，妙趣横生，读起来还真有身临其境之感了。所以，这《帝都来信》的英译本仅在法文版发行的第三年就被翻译出版，并在很短的时间内数次重印，它当时在欧洲读者们中引发的好奇之心是不难想象的。

当然，身为中国人，《帝都来信》更触动我的是，王致诚这位传教士不仅能如此敏感地解读一个遥远国度的帝皇生活，而且还如此真切地表现出对一种全然不同的异国文化的仰慕之情。相对于坦普尔爵士对中国园林艺术的情有独钟，王致诚则表现出对这皇宫御园中的园林和建筑都大加赞赏。他既能看到宫殿的中轴对称、方正严整与他自己国家的建筑同样的壮美宏阔，也被这种东方园林的多变和不规则艺术所折服，还对皇家的殿宇工程以如此不可思议的速度有序搭建而赞叹不已；他既为“中国的营建宫室之法”着迷，又清醒地认为“每个国家皆有其审美趣味及习俗”，无须要“在两者之间分出高低”。读到这些，不禁让我想起 19 世纪的福格森和他的《印度和东方建筑史》，并假想，如果当时也有条件向福格森提供在皇宫御园实地游走的机会，那么这位英国建筑史学家是否就会自然而然地把诸多偏见抛到脑后？

对个体者的叙述的确可以呈现文化交流史中最鲜活的面貌，而且，探险家，旅行者，传教士，商人或者外交使节，不同身份的人一定是因经历不同而讲述不一样的故事。若论宗教传播的成就，耶稣会士的功绩恐怕比较有限，他们在改变中国皇帝的信仰上毫无成效，他们的百般努力也没能让罗马教廷的领导者十分满意。然而，他们将早期中西文化的相遇带入到这样一个广博和深沉的交流之中，这是任何其他力量都未能企及的。没有他们，中国的历史会大不一样，而西方的历史也将重写。我们可以说，他们大都长期生活、工作在中国，甚至终其毕生，客死异乡，因此能够通过与上至帝王、下至贫民的密切接触，深入地了解中国文化、理解中国文化。我们也可以说，他们是因信仰和使命感的强大驱动，克服一切困难，将欧洲的科学技术、哲学思想和文化艺术传入中国，同时又将古老东方帝国的文化经典、典章制度、历史纪年、人文艺术、山川地貌和风土民俗介绍到西方世界，以积极的对话和知识领域的交流来消解文化的隔阂，征服帝国的信仰。而我们从《帝都来信》中还可以切实看到，一种文化交流的达成，以交流者的思想开放与文化包容为必不可少的前提。诚然，王致诚时代的欧洲自身已经处在一个知识和信仰大变革的转折期，古典的权威正在被不断打开的经验世界所挑战。但毕竟，在这位传教士成长的国度里，凡尔塞的宏伟宫殿及其几何式的园林盛景仍是当时万众钦羡的东西，而他却能以如此开放的态度从异质的东方中超越自身的传统美学，直接助长了欧洲“中国风”的兴起，同时也部分地改变了中华帝国统治者观看世界的方式。这种文化的互通与融合，在我们经过百年多的身份焦虑之后的今天再作回望，是不是意味深长呢？

文化传播总是有选择的，而要克服这交流过程中的种种障碍，更是谈何容易。这次建强完成的这个中译本，因为源自英译本，所以从一开始就已经难说它是原汁原味了。不过因为当时的英国人对中国园林的兴趣比起法国人有过之而无不及，从英文版的细读中逐句翻译，也算是对当时英国人关于遥远帝都的好奇和想象的一次体会了。建强总是毫不含糊地要以“信、达、雅”的标准完成翻译工作，但总有一些只在西方语境中才能显现其深层意义的词汇或片段，即使再流畅的中文也难以透彻表达的。比如，文中不止一次描述中国园林的不规则造景，“层层叠叠”“虽由人作”“宛若天成”，等等。而事实上，“宛若天成”这样在我们看来稀松平常的词汇，对于当时的欧洲，却是深藏在变革浪潮中的新概念。当然，这个译本最饶有趣味之处，就是它以传统意蕴的中文形式，来转达这位耶稣会传教士的游园感受，读起来有一种把这位法国人当年在帝都酿制的葡萄酒重新打开、倒入中式酒盅的感觉，喝起来真是别有滋味。

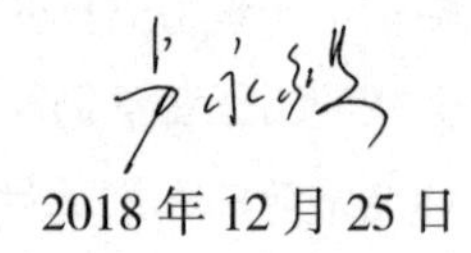

2018 年 12 月 25 日

前言

段建强

我与《帝都来信:北京皇家园林概览》的相遇纯属偶然。十年前的一日，我与恩师卢永毅教授坐在同济大学附近的一间咖啡馆内，卢老师给我看一位意大利友人 Giancarlo Floridi 赠给她的限量珂罗版 *A Particular Account of the Emperor of China's Gardens near Pekin*，我那时正跟从老师学习和研究东西方建筑和园林文化交流的历史。我翻开那泛黄的册页，这是法国耶稣会传教士王致诚 1743 年从北京写给他巴黎朋友达索先生的一封信，这位自西方远来的清宫画师，在信中详述其亲历的圆明园，令人心神往之。

我沉浸在展阅旧籍的喜悦里，仿佛两百多年前的御苑美景、林泉胜迹就在眼前。我当即就与卢老师约定，要将此信翻译出来。虽然我很快就完成了翻译初稿，但卢师鼓励我，要以文化的维度，望向历史的深处，找到一条足以让具体史料在宏阔历史中穿越时空的线索。这启发了我在后来的这些年着意于相关史料的汇集，后来也曾有意将译稿出版，以利诸位同好研究之便，但限于当时出版条件不成熟，而本人博士课业繁重，加之个人怠惰，遂将此信译稿暂时束之高阁了。

十年间，我虽然经历了诸多变迁，这份小小的译稿，却始终带在身边，得闲便拿出来翻阅，时有订正、数易其稿，自娱自乐倒也不亦乐乎。随着时间的推移，我对这封帝都来信，产生了与日俱增的浓厚兴趣。缘由此信而不断涌现的史料，似乎总如不成串的珠宝，往往散碎而不成系统。但多年后，当我重新梳理这本译稿和发现的史料，发现此信对东西方园林文化交流、欧洲英中式园林的发展以及十八世纪欧洲“中国热”的出现，竟然都发挥了独特的作用，具有提纲挈领的价值，将那些史料串联成一条熠熠生辉的珠链。于是，我又花了数年时间深入研究，从而完成《从谐奇趣到明轩：十七至二十世纪中西文化交流拾遗》的写作，正可与本书对照阅读。

非常幸运，一年前，在一个学术会议的间歇，我遇到同济大学出版社编辑吕炜女士，她专程赶来和我交流，并带给我一本《S.》，那本 J. J. 艾布拉姆斯和道格·道斯特撰写的带有诸多神秘附件的探秘小说。在畅谈中，我提起此信以及我与它的故事，她也被这封信的魅力深深吸引了：这信何尝不是一艘东西方文化交流中的“忒修斯之船”呢？历史巨轮，在时间的洪流中留落下来的不过是无数细节的碎片，而我们能够加以建构的部分，不正是去还原历史的真实么？

古训有言：“德不孤，必有邻。”不同文化之间的交流，需要历史长河中无数人为之不懈努力，才有可能出现真正汇通两种甚至多种文化的使者。这些使者不远万里，建构起跨越文化鸿沟的桥梁，而令后来者在跨文化的交流中，多些便捷、少走弯路、增进理解。耶稣会传教士们在东西方文化交流的洪流中，虽身负传教之使命，但却在不经意间以书信的方式，架构出一艘宏伟的“忒修斯之船”，而王致诚的信是这巨船之上一块小小的木头吧！我这译稿也不过是补充了上面一块小小的新木头，一如“忒修斯之船”的悖论。

翻译向来是吃力不讨好的事。在资讯发达的今日，能够促成这样一本历经十年沉睡的译稿出版，而原稿更是沉睡了二百七十余年之久，却让我倍感振奋。对此信感兴趣的学者，可径自阅读原文，我的译文权作抛砖引玉，仅供参考之便，而此信译文中所有错误，都是本人学识浅陋造成的，恳请诸位方家不吝指正。很难想象，若不是吕炜女士的恳切督促与耐心探讨，我真不知要到什么时候，才肯将这浅薄的译稿拿出来了！特此对本书责任编辑吕炜女士致以诚挚的谢忱！也感谢同济大学出版社与上海市文化基金对本书翻译、出版的具体支持。家人一直以来对我包容理解与照顾有加，一并致谢！

现今，我正在德国德累斯顿访学，也成为东西方文化交流的一员——“忒修斯之船”上一小块“新的木头”。当我坐在“易北河畔翡冷翠”的茨温格宫那敞阔的中国瓷器展廊之内，看着这数百年间东西方文化交流丰富的见证，我仍能回想起十年前那个愉快的下午，夕阳穿过橱窗照在我展阅的旧籍之上，对面卢老师眼中那殷切的期望，竟因我的怠惰拖延了这么久，实在惭愧！而这十年，虽我已毕业多年，工作生活辗转不易，但每每与老师面叙，总离不开东西方文化交流的话题，此译稿的出版，也算是给恩师交了一份小小的答卷。倘若每一位展读此卷的同好，能有我当年新发现的惊喜，我就心满意足了！

是为记。

段建强

丁酉年八月初八日，德累斯顿

目录

下面的文字是一封信的英文翻译。这封信在1743年11月1日由法国传教士王致诚（1702—1768）寄给巴黎的达索。法语原文首次出版在1749年的《耶稣会士书信集》第二十七卷第1—61页（巴黎：葛安出版）。1752年，此信由约瑟夫·斯宾塞（别名哈利·博蒙特爵士）翻译并在伦敦出版。方括号[ ]内的数字表示信件的原始页码。

The text below is a typescript of the English translation of a letter by Jean-Denis Attiret, S.J. (1702—1768) to M. d'Assaut in Paris, 1 November 1743. The French original was first published in 1749 in the *Lettres édifiantes et curieuses écrites des missions étrangères par quelques missionnaires de la compagnie de Jésus* (Paris: Guérin), 27:1-61. The following translation by Joseph Spence [alias Sir Harry Beaumont] was published in 1752 in London. The numbers in [ ] brackets indicate the original page numbers.

帝都来信
北京皇家园林概览

寄自王致诚，法国传教士。他现正受雇于中国皇帝，为其绘制那些宏伟园林的庭院。这是他写给其巴黎朋友的一封信。

译自法文
英译：哈利·博蒙特

伦敦

为在蓓尔美尔街的罗伯特·多兹利[①]刊印，由在佩特诺斯特大街[②]的 M. 库珀经销。

M.DCC.LII

① 罗伯特·多兹利（R. DODSLEY，1704—1764），英国书商、诗人、剧作家。

② 佩特诺斯特大街（Paternoster Row），伦敦城中著名的印刷与出版之街，内有朗文出版等著名机构。19 世纪，该街名几乎就是出版行业的代名词。

A

Particular Account

OF THE

EMPEROR of *CHINA's*

GARDENS

Near *PEKIN*:

IN

A LETTER from F. ATTIRET, a *French* Miſſionary, now employ'd by that Emperor to Paint the Apartments in thoſe Gardens, to his Friend at *Paris*.

Tranſlated from the *French*,

By Sir *HARRY BEAUMONT*.

LONDON:
Printed for R. DODSLEY, in *Pallmall*; and ſold by M. COOPER, in *Pater-noſter-Row*.

M.DCC.LII.

公告

大半个世纪以前，法国陆续出版了一系列传教士的信札，它们寄自遥远的世界各地。此系列已卷帙浩繁。

ADVERTISEMENT

TO THE

PUBLIC.

IT is now above half a Century, ſince the French *have been publiſhing a Collection of the Letters of their Miſſionaries; from all the moſt diſtant Parts of the World. This Collection is already grown very voluminous.*

著名的佩雷·杜赫德是这些书信制作和出版的主要推手。在他接手之前，截至1711年，仅有八卷付梓刊行；他继续这项事业，至1743年，又出版了十八卷之多。由于神父[1]辞世及其他一些原因，这一工作停滞了大约六年。直至1749年，由巴杜耶神父重新启动，并继续出版了第二十七卷。以下内容是第二十七卷的第一封信札，它与书卷中任何一封信札同样奇异动人。

① 这里指杜赫德神父（1674—1743）。

luminous. The famous Pere du Halde *was the Perſon who had the chief Hand in making and publiſhing it. There were but Eight Volumes that had appeared before he undertook the Care of it, which was in the Year* 1711; *and he carried it on, in Eighteen more, to the Year* 1743: *when the Death of that Father, and ſome other Incidents, occaſion'd an Interruption of the Work, for about Six Years. It was reſum'd in* 1749, *by F.* Patouillet; *who then publiſh'd the* 27*th Volume. The following is a Tranſlation of the Firſt Letter in that Volume; and is perhaps as curious, as any one in the whole Collection.*

一封法国传教士寄自中国的信

北京，1743年11月1日

尊贵的阁下：

非常荣幸收到您1742年10月13日和11月2日的两封来信。

A LETTER FROM A *French* Miſſionary in *China.*

Pekin; *Nov.* 1, 1743.

SIR,

IT was with the greateſt Pleaſure that I received your Two laſt Letters; one of the 13th of *October*, and the other of the 2d of *November*, 1742. I communicated the

B very

由您所寄之信札，我得以了解欧洲那些有趣的事儿，并获悉那些和我一起献身天主的传教士们的近况。我要特别向您致谢！那些装满了秸秆工艺品和花卉的箱子，已安然寄抵我处。但我恳请您今后切勿再为此破费，因为中国的这类东西远超出欧洲的水平，尤其是他们手工制作的花卉。**

* 这些主要以羽毛为原材料，染色并塑形而制作的花卉，几乎可以以假乱真，以致于经常有人情不自禁地去闻它们。或许罗马著名的“万尼曼诺夫人”（去罗马旅行的绅士们不断地将她制作的大量作品买回家），她起初正是从那些由耶稣会士寄给教皇的、来自中国的假花艺术品处习得的技艺。

* 此处省略去一两页，仅因涉及他们的私事。

very intereſting Account of the Affairs of *Europe*, which you gave me in them, to the reſt of our Miſſionaries; who join with me in our ſincere Thanks. I thank you too in particular for the Box full of Works in Straw, and Flowers, which came very ſafe to me: but I beg of you not to put yourſelf to any ſuch Expence for the future; for the *Chineſe* very much exceed the *Europeans*, in thoſe kinds of Works; and particularly, in their * Artificial Flowers †.

* Theſe are chiefly made of Feathers; colour'd, and form'd, ſo exactly like real Flowers, that one is often apt to forget one's ſelf, and ſmell to them. The famous *Signora Vannimano*, at *Rome*, (ſo many of whoſe Works in this kind are continually brought Home by our Gentlemen who travel to that City,) at firſt learn'd her Art from ſome which were ſent from *China*, by the Jeſuits; as a Preſent to the then Pope.

† Here is a Page or two omitted, as relating only to their private Affairs.

We

我们前往北京，这或许是神的使命，也可能是中国皇帝恩准的。一名官员受命与我们接洽。他竭力使我们相信，他会承担我们的开支，可最终却只是说说而已，因为几乎所有的费用，都是我们自掏腰包。一半的行程是水路，吃住皆在船上。我们不能下船登岸，甚至不得在沿途撩起舟侧舷的帘子，一探这神秘帝国的样貌，因为他们说此乃教养和礼仪，这于我们而言，实在是太古怪了。此后的一段旅程，我们更形容其如同囚笼，而他们则乐于将其称为坐轿子。白天我们在轿子里，

We came hither by the Command, or rather by the Permiſſion of the Emperor. An Officer was aſſign'd to conduct us; and they made us believe, that he would defray our Expences: but the latter was only in Words, for in Effect the Expence was almoſt wholly out of our own Pockets. Half of the Way we came by Water; and both eat, and lodg'd in our Boats: and what ſeem'd odd enough to us, was; that, by the Rules of Good-breeding received among them, we were not allow'd ever to go aſhore, or even to look out of the Windows of our Cover'd-boats to obſerve the Face of the Country, as we paſſed along. We made the latter Part of our Journey in a ſort of Cage, which they were

B 2 pleas'd

终日被禁言。到了晚上，我们又被带到客栈（那是多么糟糕的客栈啊）。我们在整个行程中没有获得更多的信息，甚至对这个国家仍一无所知，还不如在房间里安静地睡大觉。就是这样，满怀着未被满足的好奇心，我们来到了京城。

他们所言非虚，我们先前穿越的，乃是这一帝国破败不堪的地区，尽管行程接近两千英里，漫漫路途之中并没有太多值得去注意的事物，甚至没有任何纪念碑或宏伟的建筑物。所谓价值与美，除了存在于一些为神像而修建的寺庙和一层楼高的木构建筑外，似乎只存在于其中糟糕的绘画和稀松平常的粉饰中。

pleas'd to call a Litter. In this too we were ſhut up, all Day long; and at Night, carried into our Inns; (and very wretched Inns they are!) and thus we got to *Pekin*; with our Curioſity quite unſatisfy'd, and with ſeeing but very little more of the Country, than if one had been ſhut up all the while in one's own Chamber.

Indeed they ſay, that the Country we paſſed is but a bad Country; and that, tho' the Journey is near 2000 Miles, there is but little to be met with on the Way that might deſerve much Attention: not even any Monuments, or Buildings, except ſome Temples for their Idols; and thoſe built of Wood, and but one Story high: the chief Value and

Beauty

其实，任何一个见识过法国或意大利建筑的人，对世界另一端的这些所见，都会有味同嚼蜡之感。

然而我必须指出的是，帝都北京的皇宫御苑则不然，无论是其设计还是竣工后的效果，其中的每一样事物均皆宏大壮美。我被深深吸引，因为此前我在世界其他地方都未曾见识过与之相类似的。

Beauty of which ſeem'd to conſiſt in ſome bad Paintings, and very indifferent Varniſh-works. Indeed any one that is juſt come from ſeeing the Buildings in *France* and *Italy*, is apt to have but little Taſte, or Attention, for whatever he may meet with in the other Parts of the World.

However I muſt except out of this Rule, the Palace of the Emperor of *Pekin*, and his Pleaſure-houſes; for in them every thing is truly great and beautiful, both as to the Deſign and the Execution: and they ſtruck me the more, becauſe I had never ſeen any thing that bore any manner of Reſemblance to them, in any Part of the World that I had been in before.

I ſhould

如若通过我的描述，会令您对帝都的皇宫御苑有所体悟，我将备感荣幸。但我也自知那几乎是不可能的，因为这里的一切，与我们所熟知的建筑风格与建造法式全然不同。唯一能感受它的方式，就是身临其境。若能不计时间，我决意倾力将其逐一描绘下来，并将画稿寄往欧洲。

这座宫苑[①]，至少像第戎*一样大；如此，您便能够通过对比，认识到其壮美。

* 一个美丽的法国城市；勃艮第大区首府，周长三至四英里。

① 译者注：这里的 The Palace 指当时正在建设中的圆明园。

I ſhould be very glad, if I could make ſuch a Deſcription of theſe, as would give you any juſt Idea of them; but that is almoſt impoſſible; becauſe there is nothing in the Whole, which has any Likeneſs to our manner of Building, or our Rules of Architecture. The only way to conceive what they are, is to ſee them: and if I can get any time, I am reſolved to draw ſome Parts of them as exactly as I can, and ſend them into *Europe*.

The Palace is, at leaſt, as big as * *Dijon*; which City I chuſe to name to you, becauſe you are ſo well ac-

* A handſome City in *France*; and the Capital one, in the Province of *Burgundy*: between Three and Four Miles round.

quainted

这座宫苑由众多形态各异的建筑组合而成，虽然互相分离，却呈现出整体的均衡与美感。建筑被巨大的庭园、葱郁的林地和满植奇花异草的花园所分隔。所有建筑物主立面上的构件都做了鎏金，或施以清漆，或饰以彩绘，耀人眼目。建筑内部则饰以众多精美绝伦且价值连城的陈设，大多都是从中国各地、印度，甚至从欧洲搜寻所得的至宝。

宫苑着实迷人。它占地甚广，堆山造势，高度从二十英尺至六十英尺不等，其间沟壑纵横，不计其数。

quainted with it. This Palace consists of a great Number of different Pieces of Building; detach'd from one another, but disposed with a great deal of Symmetry and Beauty. They are separated from one another by vast Courts, Plantations of Trees, and Flower-gardens. The principal Front of all these Buildings shines with Gilding, Varnish-work, and Paintings; and the Inside is furnish'd and adorn'd with all the most beautiful and valuable Things that could be got in *China*, the *Indies*, and even from *Europe*.

As for the Pleasure-houses, they are really charming. They stand in a vast Compass of Ground. They have raised Hills, from 20 to 60 Foot high; which form a great Number

山脚之下，清泉潺潺不息，汇流形成较大的水面和湖泊。他们乘坐华丽的舟船，穿行于这些溪流、湖泊与河流。我就曾见过一艘七十八英尺长、二十四英尺宽的游船，船上修造着一座堂皇气派的宫室。每处山谷中皆有临岸屋宇，具备协调、各臻其妙，有庭院、敞廊、暗廊、花圃、泉瀑之属，至此而全胜一览，足称美景也。

自山谷出，往赴他谷，不若欧陆林荫宽道平直，

Number of little Valleys between them. The Bottoms of theſe Valleys are water'd with clear Streams; which run on till they join together, and form larger Pieces of Water and Lakes. They paſs theſe Streams, Lakes, and Rivers, in beautiful and magnificent Boats. I have ſeen one, in particular, 78 Foot long, and 24 Foot broad; with a very handſome Houſe raiſed upon it. In each of theſe Valleys, there are Houſes about the Banks of the Water; very well diſpoſed: with their different Courts, open and cloſe Porticos, Parterres, Gardens, and Caſcades: which, when view'd all together, have an admirable Effect upon the Eye.

They go from one of the Valleys to another, not by formal ſtrait Walks

而是以曲折迂回之小径通之，并有小巧亭台、叠石假山石洞点缀于曲径侧旁。每一处谷涧境皆独辟，或因地制宜，或随建筑形制而布局迥异。

逶迤伸展的山丘之上遍植林木，尤以开花树木为多，皆为常见的品种。无论运河之驳岸，还是涧流之侧旁，均不用光滑的石材饰面，亦非呈直线形态（一如我们欧洲园林所为），而是以不同大小的天然岩石堆叠出岸线，或前凸或后退，犬牙交错，饶具野趣；众多艺术品散置其间，颇有匠心，宛若天成。在一些

Walks as in *Europe*; but by various Turnings and Windings, adorn'd on the Sides with little Pavilions and charming Grottos: and each of theſe Valleys is diverſify'd from all the reſt, both by their manner of laying out the Ground, and in the Structure and Diſpoſition of its Buildings.

All the Riſings and Hills are ſprinkled with Trees; and particularly with Flowering-trees, which are here very common. The Sides of the Canals, or leſſer Streams, are not faced, (as they are with us,) with ſmooth Stone, and in a ſtrait Line; but look rude and ruſtic, with different Pieces of Rock, ſome of which jut out, and others recede inwards; and are placed with ſo much Art, that you would take it to be the

C Work

区域，水面相当辽阔，另一些则窄如蛇行；此处蜿蜒折曲，彼处漫游铺展，好似被山与石挤迫推进而成。水岸上遍植花草，甚至岩隙间也花草葱茏，好像它们早就在那里自然生长一般。花草品种丰富，且随时令与季节而易，变化多姿。

在这些溪流之上，总有铺着小石子的散步小径行于山谷之间。这些路径，或近水岸，或稍远离之，皆具自由蜿蜒之意。

Work of Nature. In ſome Parts the Water is wide, in others narrow; here it ſerpentizes, and there ſpreads away, as if it was really puſh'd off by the Hills and Rocks. The Banks are ſprinkled with Flowers; which riſe up even thro' the Hollows in the Rock-work, as if they had been produced there naturally. They have a great Variety of them, for every Seaſon of the Year.

Beyond theſe Streams there are always Walks, or rather Paths, pav'd with ſmall Stones; which lead from one Valley to another. Theſe Paths too are irregular; and ſometimes wind along the Banks of the Water, and at others run out wide from them.

On

每个山谷入口，皆可见屋宇于前。正面柱廊周匝，柱间有窗。凡间架木构均鎏金、彩绘，或施以清漆。屋顶覆瓦，色彩纷呈：红、黄、蓝、绿、紫，各色按一定比例法式，间杂铺设，形成丰富而协调的变化层次，极尽美观。几乎所有屋宇皆仅一层，屋基自地面起，屋高二至八英尺。若要拾级而上，登堂入室，并没有规则的石阶，取而代之的是一组粗犷的堆石，层层叠叠，虽由人作，宛自天开。

On your Entrance into each Valley, you ſee its Buildings before you. All the Front is a Colonnade, with Windows between the Pillars. The Wood-work is gilded, painted, and varniſh'd. The Roofs too are cover'd with varniſh'd Tiles of different Colours; Red, Yellow, Blue, Green, and Purple: which by their proper Mixtures, and their manner of placeing them, form an agreeable Variety of Compartiments and Deſigns. Almoſt all theſe Buildings are only one Story high; and their Floors are raiſed from Two to Eight Foot above the Ground. You go up to them, not by regular Stone Steps, but by a rough Sort of Rock-work; form'd as if there had been ſo many Steps produced there by Nature.

宫室之内完美诠释了何为华丽，不仅布局合理，家私与陈设亦皆典雅珍贵。庭院廊庑之间，您会见到插满鲜花的黄铜花瓶、瓷花瓶和大理石花瓶；阶前陈列的，不是裸体雕像，而是若干瑞兽神像；在大理石台基上，摆着焚香之鼎。

如我前言所述，每一山谷皆设一宫殿。就其占地而言，相对于整座宫苑，固然不甚大，然亦不算小，足可接纳下欧洲最显要的贵族及其所有随从仍有余。营造

The Inſide of the Apartments anſwers perfectly to their Magnificence without. Beſide their being very well diſpoſed, the Furniture and Ornaments are very rich, and of an exquiſite Taſte. In the Courts, and Paſſages, you ſee Vaſes of Braſs, Porcelain, and Marble, fill'd with Flowers: and before ſome of theſe Houſes, inſtead of naked Statues, they have ſeveral of their Hieroglyphical Figures of Animals, and Urns with Perfumes burning in them, placed upon Pedeſtals of Marble.

Every Valley, as I told you before, has it's Pleaſure-houſe: ſmall indeed, in reſpect to the whole Incloſure; but yet large enough to be capable of receiving the greateſt Noble-

宫室所用木材，有用松木者，斥以巨资自一千五百英里之外运至此处。您能想象得出在这宫苑之中，所有的谷壑之间，有多少座类似的宫殿么？有超过两百处，且内监的官舍尚不计入其内。每座宫殿皆由专属的内监打理，内监的住舍常就近营建，一般来说，距离宫殿不会超过五至六英尺。这些内监住舍形制非常朴素，因此会借助一些影壁或假山隐蔽起来。

Nobleman in *Europe*, with all his Retinue. Several of theſe Houſes are built of Cedar; which they bring, with great Expence, at the Diſtance of 1500 Miles from this Place. And now how many of theſe Palaces do you think there may be, in all the Valleys of the Incloſure? There are above 200 of them: without reckoning as many other Houſes for the Eunuchs; for they are the Perſons who have the Care of each Palace, and their Houſes are always juſt by them; generally, at no more than Five or Six Foot Diſtance. Theſe Houſes of the Eunuchs are very plain: and for that Reaſon are always concealed, either by ſome Projection of the Walls, or by the Interpoſition of their artificial Hills.

Over

涧流之上皆设桥，桥间距适当，以便于各处通达。这些桥梁多以砖砌筑，或以天然岩石堆叠，亦有用木材者。桥面必抬高，以便船只在下穿行。桥以汉白玉为栏杆，皆精雕细琢，千姿百态，无论是图样还是造型皆各不相同。

不要理所当然地认为，这些桥像我们国家那样是笔直的。恰恰相反，它们回环绕行，蜿蜒曲折，毫无规

Over the running Streams there are Bridges, at proper Diſtances, to make the more eaſy Communication from one Place to another. Theſe are moſt commonly either of Brick or Free-ſtone, and ſometimes of Wood; but are all raiſed high enough for the Boats to paſs conveniently under them. They are fenced with Balliſters finely wrought, and adorned with Works in Relievo; but all of them varied from one another, both in their Ornaments, and Deſign.

Do not imagine to yourſelf, that theſe Bridges run on, like ours, in ſtrait Lines: on the contrary, they generally wind about and ſerpentize to ſuch a Degree, that ſome of them, which, if they went on regularly, would

律可言。一些桥如是直的，则长度仅需三十英尺或四十英尺即可，但竟会因曲折而延伸至一百英尺或两百英尺。有时桥上可见供人休憩的凉亭（时而位于中段，时而位于尽端）。凉亭有四柱、八柱，甚或十六柱者。桥亭之选址，往往可坐享最迷人的美景。有些桥会在两端设凯旋拱门，或以木构或以白色大理石构，皆形制优美，但迥异于我在欧洲所见过的任何事物。

我已告诉过您，那些小小的涧溪或河流最终汇成了

would be no more than 30 or 40 Foot long, turn ſo often and ſo much as to make their whole Length 100 or 200 Foot. You ſee ſome of them which, (either in the Midſt, or at their Ends,) have little Pavilions for People to reſt themſelves in; ſupported ſometimes by Four, ſometimes by Eight, and ſometimes by Sixteen Columns. They are uſually on ſuch of the Bridges, as afford the moſt engaging Proſpects. At the Ends of other of the Bridges there are triumphal Arches, either of Wood, or white Marble; form'd in a very pretty Manner, but very different from any thing that I have ever ſeen in *Europe*.

I have already told you, that theſe little Streams, or Rivers, are carried

3 on

几个广阔的大池与湖泊。其中之一，周长接近五英里，他们称之为“海子”或“海”，是整个宫苑中的绝美之处。湖岸上几处宫殿，则被假山溪流分隔开来，各得其所。

最值得赏鉴的，是海子中的一座岛屿及岩石。它高出水面约六英尺，嶙峋质朴，野趣横生。在这石岛之上，有一小小的宫殿。虽称其为“小”，实则有屋百间以上。其四面临湖，华美精妙，佳境难以言传，令人心驰

on to ſupply ſeveral larger Pieces of Water, and Lakes. One of theſe Lakes is very near Five Miles round; and they call it a Meer, or Sea. This is one of the moſt beautiful Parts in the whole Pleaſure-ground. On the Banks, are ſeveral Pieces of Building; ſeparated from each other by the Rivulets, and artificial Hills above-mentioned.

But what is the moſt charming Thing of all, is an Iſland or Rock in the Middle of this Sea; rais'd, in a natural and ruſtic Manner, about Six Foot above the Surface of the Water. On this Rock there is a little Palace; which however contains an hundred different Apartments. It has Four Fronts; and is built with inexpreſſible Beauty and

神往。此处，可尽得山水之妙趣，可遍览散落海子周围的所有宫殿。山皆行止于此，水皆缘起于斯、汇聚于斯，而桥梁、亭台、拱廊以及宫殿间隔物障景的林木，也都在此一览无余。

迷人的海子周围景致变幻莫测，没有两处是类似的。

and Tafte; the Sight of it ftrikes one with Admiration. From it you have a View of all the Palaces, fcattered at proper Diftances round the Shores of this Sea; all the Hills, that terminate about it; all the Rivulets, which tend thither, either to difcharge their Waters into it, or to receive them from it; all the Bridges, either at the Mouths or Ends of thefe Rivulets; all the Pavilions, and Triumphal Arches, that adorn any of thefe Bridges; and all the Groves, that are planted to feparate and fcreen the different Palaces, and to prevent the Inhabitants of them from being overlooked by one another.

The Banks of this charming Water are infinitely varied: there are no two Parts of it alike. Here you fee

D Keys

往这边看，环湖的宫殿，或有长廊，或有林荫小径，可以拾级而下，来到铺满平整石块的水边；往那边看，又是另一番匠心独具、嶙峋交错的石景；此处，自然地貌的每层台地都由曲径通联，登高而上，宫殿层层叠叠，甚或可围合成一露天剧场；彼处，乱花迷眼，前驱而行，得见移自远山深谷的参天古树，根深叶茂，亦有异国珍植，奇花异珍果馔之属，无一不备。

Keys of ſmooth Stone; with Porticoes, Walks, and Paths, running down to them from the Palaces that ſurround the Lake: there, others of Rock-work; that fall into Steps, contrived with the greateſt Art that can be conceived: here, natural Terraces with winding Steps at each End, to go up to the Palaces that are built upon them; and above theſe, other Terraces, and other Palaces, that riſe higher and higher, and form a ſort of Amphitheatre. There again a Grove of Flowering-trees preſents itſelf to your Eye; and a little farther, you ſee a Spread of wild Foreſt-trees, and ſuch as grow only on the moſt barren Mountains: then, perhaps, vaſt Timber-trees with their Under-wood; then, Trees from all foreign Countries; and then, ſome

all

滨湖池岸，复有珍禽异兽之笼毚无数。各类水禽半入水中半居池岸，陆兽则远离水岸，围以栏篱圈养，甚至还配有小型的狩猎场。但在所有动物中，被中国人视为珍品的，是一种鱼[①]，鱼身大部分灿若黄金，亦有银色的，或是赤、青、蓝、绛、黑或诸色混合者。

① 这种鱼应为锦鲤。

all blooming with Flowers, and others all laden with Fruits of different Kinds.

There are alſo on the Banks of this Lake, a great Number of Net-work-houſes, and Pavilions; half on the Land, and half running into the Lake, for all ſorts of Water-fowl: as farther on upon the Shore, you meet frequently with Menageries for different ſorts of Creatures; and even little Parks, for the Chace. But of all this ſort of Things, the *Chineſe* are moſt particularly fond of a kind of Fiſh, the greater Part of which are of a Colour as brilliant as Gold; others, of a Silver Colour; and others of different Shades of Red, Green, Blue, Purple, and Black: and ſome, of all Sorts of Colours mixt together.

D 2 There

宫苑各处鱼沼甚多，但在此湖中鱼最多。鱼沼需水域广阔，且需用细铜丝编成网栏围护，网栏开口细密，以防鱼游至全池。

如要领略到这片湖面的盛景实况，我想您应该在游船环集之时来此游览。水面或波光粼粼、或平静如镜，时或泛舟，时或垂钓，时或对阵*，时或竞舟，还有其他

* 我曾在我们这里，准确些描述，是在法国莱昂，见过这样的水上阵仗。对战双方均会派出一人，尽可能稳定地立于两艘船的船头，用其左手执盾，右手执矛。每艘船都有人数相当的船员，竭尽全力驾驭船舰前冲。两个战斗的前锋，会用自己的矛攻击对方，但通常对战双方，或至少是一方，会因为受到撞击，或跌入船中，或跌落水中（经常发生）。掉入水中这样的结局就是这类古怪娱乐消遣中最精彩的环节。

There are ſeveral Reſervoirs for theſe Fiſh, in all Parts of the Garden; but the moſt conſiderable of them all is at this Lake. It takes up a very large Space; and is all ſurrounded with a Lattice-work of Braſs-wire: in which the Openings are ſo very fine and ſmall, as to prevent the Fiſh from wandering into the main Waters.

To let you ſee the Beauty of this charming Spot in its greateſt Perfection, I ſhould wiſh to have you tranſported hither when the Lake is all cover'd with Boats; either gilt, or varniſh'd: as it is ſometimes, for taking the Air; ſometimes, for Fiſhing; and ſometimes, for * Juſts, and Combats,

* I have ſeen of this ſort of Juſts upon the Water, in our Parts of the World; and particularly, at *Lions* in *France*. The Champions ſtand, as firmly as they are able, on the Prows of two Boats;

and

娱乐，皆在水面上开展。但至为重要的，是在一些晴朗的夜晚，烟火盛宴会在此举行。当是时，园内殿宇齐明，舟船耀目，火树银花。盖因中国烟花技术胜我国良多，其精彩为所未见。仅就在下个人囿见，中国烟花较之欧陆，无论法国或意大利，皆可完胜之。

帝后妃嫔*与侍奉他们的内监宫娥起居生活之地，

* 原文为“les Koucifeys, les Feys, les Pines, les Kouci-gins, et les Tchangtsays:”并且专门做了注解，这些都是对皇帝不同等级的嫔妃们的尊称①，表示这些嫔妃的尊贵之别以及皇帝对她们的喜爱程度。我以为不值得渴求这些繁文缛节的名头，此处可以忽略它而采用笼统的称呼。

① 对应清代后宫，分别为贵妃、妃、嫔、贵人、常在。

and other Diverſions, upon the Water: but above all, on ſome fine Night, when the Fire-works are play'd off there; at which time they have Illuminations in all the Palaces, all the Boats, and almoſt on every Tree. The *Chineſe* exceed us extremely in their Fire-works: and I have never ſeen any thing of that Kind, either in *France* or *Italy*, that can bear any Compariſon with theirs.

The Part in which the Emperor uſually reſides here, with the Em-

with a Shield in their left Hands, and a blunted Spear in their Right. There is an equal Number of Rowers in each of the Boats, who drive them on with a great deal of Impetuoſity. The two Combatants charge each other with their Spears: and often both, but almoſt always one or other of them, is driven backward on the Shock; either down into his Boat, or (which often happens) into the Water: which latter makes one of the principal Parts in this odd ſort of Diverſion.

preſs.

乃大量殿舍园囿之集成，有若城池。在规模上，至少和我们的多尔市*一样大。其他各处宫殿，则仅用于游赏及偶尔会饮之用。

皇帝起居之所，临近宫苑正门，前有正殿，其后为

* 多尔市，法国孔泰区第二大城市。

press, his * favourite Mistresses, and the Eunuchs that attend them, is a vast Collection of Buildings, Courts, and Gardens; and looks itself like a City. 'Tis, at least, as big as our City † of *Dole*. The greater Part of the other Palaces is only used for his walking; or to dine or sup in, upon Occasion.

This Palace for the usual Residence of the Emperor is just within the grand Gate of the Pleasure-ground. First are the Ante-cham-

* The Original says; "les *Koucifeys*, les *Feys*, "les *Pines*, les *Kouci-gins*, et les *Tchangtsays*:" and informs us in a Note, that these are so many different Titles of Honour, for the different Classes of such of the Emperor's Mistresses, as are most in his Favour. I did not think it worth while to set down all these hard Names in the Text; and, perhaps, they might as well have been omitted even here.

† The second City for Size in the *Franche Comté*.

bers;

朝寝，再后则是庭院和园林。内宫整体上就是一座岛屿，周围被既宽又深的护城河环绕。这是一种类似“塞拉利奥”[①]的建筑。在不同的居所之内，您可看到所有能想象到的最美好的东西（我指的是那些具有浓郁中国风情的各种东西），家具、陈设、绘画，名贵的木器，中国和日本的漆器，古瓷花瓶，绫罗绸缎，黄金白银，不胜枚举。这些珍品汇集于此，极尽华美，可谓天然物产与人工巧思并萃杂陈。

从帝宫起始，筑有通衢，达至小城。小城位于整座

① 塞拉利奥（Seraglio）指土耳其苏丹宫殿中妻妾的闺房。

bers; then, the Halls for Audience: and then, the Courts, and Gardens belonging to them. The Whole forms an Iſland; which is entirely ſurrounded by a large and deep Canal. 'Tis a ſort of Seraglio; in the different Apartments of which you ſee all the moſt beautiful things that can be imagin'd, as to Furniture, Ornaments, and Paintings, (I mean, of thoſe in the *Chineſe* Taſte;) the moſt valuable Sorts of Wood; varniſh'd Works, of China and Japan; antient Vaſes of Porcelain; Silks, and Cloth of Gold and Silver. They have there brought together, all that Art and good Taſte could add to the Riches of Nature.

From this Palace of the Emperor a Road, which is almoſt ſtrait, leads you

宫苑中部，城中有广场，每侧长约一英里，四向皆设城门，以辨方正位；有城楼、墙垣、堞雉、护栏和城垛；大道广场、庙宇市廛、商店衙署、宫殿船埠，无所不备。真可谓北京都城之缩影。

您想必会问，营建此城有什么用意吗？它是皇帝在遇到叛乱或革命时撤退的安全之地吗？它很有可能是为这一目的而营造的，抑或极有可能反映出设计者的另一种构思：即，当皇帝想体察民生时，能够立即在

you to a little Town in the Midſt of the whole Incloſure. 'Tis ſquare; and each Side is near a Mile long. It has Four Gates, anſwering the Four principal Points of the Compaſs; with Towers, Walls, Parapets, and Battlements. It has it's Streets, Squares, Temples, Exchanges, Markets, Shops, Tribunals, Palaces, and a Port for Veſſels. In one Word, every thing that is at *Pekin* in Large, is there repreſented in Miniature.

You will certainly aſk, for what Uſe this City was intended? Is it that the Emperor may retreat to it as a Place of Safety, on any Revolt, or Revolution? It might indeed ſerve well enough for that Purpoſe; and poſſibly that Thought had a Share in the

这座具体而微的城池中体验到伟大城市中的所有繁华和忙碌的乐趣。

中国的皇帝为彰显其威严而倍受约束，他甚至不能直接向人民展现自己的仪容。即使当他出宫巡察，他对自己所经之地也看不到什么。因其出巡沿途的商业市贾、集镇民居均需闭户，各处施以屏障，到处张贴布榜，以使臣民不得亲见之。出巡之前几小时，必先清道、禁绝人行，有误闯侵道者，必将由其禁卫军严厉处置。

the Mind of the Perſon, who at firſt deſign'd it: but it's principal End was to procure the Emperor the Pleaſure of ſeeing all the Buſtle and Hurry of a great City in little, whenever he might have a Mind for that ſort of Diverſion.

The Emperor of *China* is too much a Slave to his Grandeur ever to ſhew himſelf to his People, even when he goes out of his Palace. He too ſees nothing of the Town, which he paſſes thorough. All the Doors and Windows are ſhut up. They ſpread wide Pieces of Cloth every where, that no body may ſee him. Several Hours before he is to paſs through any Street, the People are forewarned of it; and if any ſhould be found there whilſt he paſſes, they would be

E handled

无论何时出巡，皆有骈骑开道，森列道路两侧，既为安全计，亦可阻防他人入内。作为中华帝国的皇帝，他们被迫生活在这种怪异的孤独之中，他们只能一直倾力弥补自己在公共娱乐活动中的缺位（因其位分尊贵，绝无亲民之可能），不得不别开生面，根据他们不同的口味和幻想，聊以自娱。

在这两朝中[①]（正是因为当朝清帝的父皇下令营建了这座小城），每年总会有那么几次，小城被拨用作

① 指清雍正、乾隆两朝。

handled very ſeverely by his Guards. Whenever he goes into the Country, two Bodies of Horſe advance a good Way before him, on each Side of the Road; both for his Security, and to keep the Way clear from all other Paſſengers. As the Emperors of *China* find themſelves obliged to live in this ſtrange ſort of Solitude, they have always endeavoured to ſupply the Loſs of all public Diverſions, (which their high Station will not ſuffer them to partake,) by ſome other Means or Inventions, according to their different Taſtes and Fancies.

This Town therefore, in theſe Two laſt Reigns, (for it was this Emperor's Father who order'd it to be built,) has been appropriated for the

为表演场。宫监入城，乔充商贩市贾，商业、市侩、手艺人等，应有尽有，熙攘往来其中，俨然帝都市廛。在指定的时间，每个太监都穿戴上指派给他的服装，或为工役、或为士卒，推车担筐……简言之，各有其事，各司其职。开埠迎船，商铺开张，陈肆列货，各分路段；四分之一的人贩卖丝绸，余下的人卖棉布；另一条街则贩卖瓷器，还有一条街贩卖漆器。您能买到任何想要的

the Eunuchs to act in it, at ſeveral times in the Year, all the Commerce, Marketings, Arts, Trades, Buſtle, and Hurry, and even all the Rogueries, uſual in great Cities. At the appointed Times, each Eunuch puts on the Dreſs of the Profeſſion or Part which is aſſigned to him. One is a Shopkeeper, and another an Artiſan; this is an Officer, and that a common Soldier: one has a Wheel-barrow given him, to drive about the Streets; another, as a Porter, carries a Baſket on his Shoulders. In a word, every one has the diſtinguiſhing Mark of his Employment. The Veſſels arrive at the Port; the Shops are open'd; and the Goods are expoſed for Sale. There is one Quarter for thoſe who ſell Silks, and another for thoſe who ſell Cloth; one Street for Porcelain, and another

E 2 for

东西，木器衣装、妇女珍饰、旧籍典藏、古玩雅好，应有尽有。酒肆茶坊、行台村店开张，果酱走贩、针线游商等皆沿街吆喝，揽裾而售，均无禁止。而皇帝微服临幸之时，亦与其臣民鲜有区别。每个人为了完成交易任务，都竭尽全力，甚或破口大骂，拳脚相加。集市中的

for Varnish-works. You may be supply'd with whatever you want. This Man sells Furniture of all sorts; that, Cloaths and Ornaments for the Ladies: and a third has all kinds of Books, for the Learned and Curious. There are Coffee-houses too, and Taverns, of all Sorts, good and bad: beside a Number of People that cry different Fruits about the Streets, and a great Variety of refreshing Liquors. The Mercers, as you pass their Shops, catch you by the Sleeve; and press you to buy some of their Goods. 'Tis all a Place of Liberty and Licence; and you can scarce distinguish the Emperor himself, from the meanest of his Subjects. Every body bauls out what he has to sell; some quarrel, others fight: and you have all the Confusion of a Fair

about

混乱之象，真假难辨，活灵活现。更有衙役押解闹事者，对簿公堂，申明王法，虽纯属虚构，但却真施以杖责，立即执行，悉以游戏出之，皆在取悦君王。皇帝的这项消遣，时常会让他的演员们饱受皮肉之苦。

在这样的表演中，甚至不忘记“神偷”的角色。一批最轻巧灵活的太监被择以担此重任，若一旦罪行败露，他们会被当众羞辱（至少他们会遭受形式上的谴责），依据罪行

about you. The public Officers come and arreſt the Quarrellers; carry them before the Judges, in the Courts for Juſtice; the Cauſe is try'd in form; the Offender condemn'd to be baſtinado'd; and the Sentence is put in Execution: and that ſo effectually, that the Diverſion of the Emperor ſometimes coſts the poor Actor a great deal of real Pain.

The Myſtery of Thieving is not forgot, in this general Repreſentation. That noble Employ is aſſign'd to a conſiderable Number of the clevereſt Eunuchs; who perform their Parts admirably well. If any one of them is caught in the Fact, he is brought to Shame; and condemn'd, (at leaſt they go through the Form of condemning him,) to be ſtigma-

2 tiz'd,

程度与性质被当场杖责，甚至会被刺青发配。若他们偷窃得巧妙，就能笑到最后，欢呼雀跃，受到大家的掌声喝彩，而受害者则呼告无门。然而，在市集结束后，所窃之物仍需物归原主。

此类市集（如我之前所言），仅为取悦帝后妃嫔而设，王公显贵鲜有获准参与这非同寻常的仪式者，偶有参与，也必须等到嫔妃们玩累回宫之后。市集中展陈与交易的商货，主要归京城商贾所有，商贾委托内监代售，

tiz'd, baſtinado'd, or baniſh'd; according to the Heinouſneſs of the Crime, and the Nature of the Theft. If they ſteal cleverly, they have the Laugh on their Side; they are applauded, and the Sufferer is without Redreſs. However, at the End of the Fair, every thing of this Kind is reſtor'd to the proper Owner.

This Fair, (as I told you before,) is kept only for the Entertainment of the Emperor, the Empreſs, and his Miſtreſſes. 'Tis very unuſual for any of the Princes, or Grandees, to be admitted to ſee it: and when any have that Favour, it is not till after the Women are all retired to their ſeveral Apartments. The Goods which are expos'd and ſold here, belong chiefly to the Merchants of *Pekin*;

所有交易真实可查，因此市集中的讨价还价并非假装演戏。尤其是皇帝本人总是买很多东西，由此可以确定，商贾们经常能在皇帝身上大捞一笔。一些宫女也会讨价还价，而太监们竟不示弱。所有的交易若不是假中存真，就会过于枯燥，一本正经，现在则在熙攘间多出许多热闹和趣味。

有时，市集之后还会做农事。在宫苑中辟出一隅之地，

Pekin; who put them into the Hands of the Eunuchs, to be ſold in reality: ſo that the Bargains here are far from being all pretended ones. In particular, the Emperor himſelf always buys a great many things; and you may be ſure, they ask him enough for them. Several of the Ladies too make their Bargains; and ſo do ſome of the Eunuchs. All this trafficking, if there was nothing of real mixt with it, would want a great deal of that Earneſtneſs and Life, which now make the Buſtle the more active, and the Diverſion it gives the greater.

To this Scene of Commerce, ſometimes ſucceeds a very different one; that of Agriculture. There is a Quarter, within the ſame Incloſure,

划为专区，从事农作。在此您可以看到田地、草场、农庄与四处散布的小茅舍，牛、犁以及农耕所必需的生产用品齐备。他们于此处播种麦、稻、黍和其他各种谷物。年谷丰登，他们从土地中获得回报。总之，他们在此模仿一切乡土民情，竭尽所能，一事一物，皆显农村质朴，一举一动，皆合民俗。

无疑您应该对中国著名的盛会——元宵节有所耳闻。每年的正月十五皆要欢庆佳节。是日，无论贫富，

closure, which is set apart for this Purpose. There you see Fields, Meadows, Farm-houses, and little scatter'd Cottages; with Oxen, Ploughs, and all the Necessaries for Husbandry. There they sow Wheat, Rice, Pulse, and all other sorts of Grain. They make their Harvest; and carry in the Produce of their Grounds. In a Word, they here imitate every thing that is done in the Country; and in every thing express a rural Simplicity, and all the plain Manners of a Country Life, as nearly as they possibly can.

Doubtless you have read of the famous Feast in *China*, call'd *The Feast of the Lanthorns*. It is always celebrated on the 15th Day of the first Month. There is no *Chinese* so poor

皆亮灯为乐。灯具花式繁多，大小不一，价分贵贱。元宵盛会之时，举国灯火通明，尤以皇宫中为最，而宫苑中至为美观，容我向您一一描述。彩灯高悬于殿堂廊庑之上，船灯则在水溪池沼中随波逐流。假山与桥梁之上，林木与花树之间，各式灯具巧制玲珑，状如飞禽走兽，花、果、瓶、盆、船艇诸式咸备。更有以丝绸、

poor, but that upon this Day he lights up his Lanthorn. They have of them of all ſorts of Figures, Sizes, and Prices. On that Day, all *China* is illuminated: but the fineſt Illuminations of all are in the Emperor's Palaces; and particularly in theſe Pleaſure-grounds, which I have been deſcribing to you. There is not a Chamber, Hall, or Portico, in them, which has not ſeveral of theſe Lanthorns hanging from the Cielings. There are ſeveral upon all the Rivulets, Rivers, and Lakes; made in the Shape of little Boats, which the Waters carry backward and forward. There are ſome upon all the Hills and Bridges, and almoſt upon all the Trees. Theſe are wrought mighty prettily, in the Shapes of different Fiſhes, Birds, and Beaſts; Vaſes, Fruits, Flowers; and Boats of different Sorts

F and

牛角、玻璃、母贝等，材质不下千类，亦有饰彩绘刺绣者，不胜枚举，价格各异。有价值千金者，形式繁复，工艺精湛，颇具巧思，美轮美奂，实难尽述之。华人于此及建筑营构之艺匠，实非吾国所能及也。

他们视其宫室建筑形制为理所当然，对吾国建筑知之

and Sizes. Some are made of Silk; ſome of Horn, Glaſs, Mother of Pearl, and a thouſand other Materials. Some of them are painted; others embroider'd; and of very different Prices. I have ſeen ſome of them which could never have been made for a thouſand Crowns. It would be an endleſs thing, to endeavour to give you a particular Account of all their Forms, Materials, and Ornaments. It is in theſe, and in the great Variety which the *Chineſe* ſhew in their Buildings, that I admire the Fruitfulneſs of their Invention; and am almoſt tempted to own, that we are quite poor and barren in Compariſon of them.

Their Eyes are ſo accuſtom'd to their own Architecture, that they

2 have

甚少。我可否告知阁下，当我展示我国最著名的建筑图片时他们的言论？我国宫殿之高大宏阔，令他们惊讶异常。他们视我国的街道是挖空崇山峻岭才成的通道，视我国房屋为充满了熊兽洞穴的冲天巨石，更视我国之层叠高楼为不可理喻。他们无法想象，我们怎么能每天冒着摔断脖子的危险，攀爬到四五层高去活动。“毫无疑问，”清康熙皇帝尝言（虽然他翻阅了我们欧洲的一些屋

have very little Tafte for ours. May I tell you what they fay when they fpeak of it, or when they are looking over the Prints of fome of our moft celebrated Buildings? The Height and Thicknefs of our Palaces amazes them. They look upon our Streets, as fo many Ways hollowed into terrible Mountains; and upon our Houfes, as Rocks pointing up in the Air, and full of Holes like Dens of Bears and other wild Beafts. Above all, our different Stories, piled up fo high one above another, feem quite intolerable to them: and they cannot conceive, how we can bear to run the Rifk of breaking our Necks, fo commonly, in going up fuch a Number of Steps as is neceffary to climb up to the Fourth and Fifth Floors. "Undoubtedly, (faid the

F 2 "Em-

宇设计），“欧罗巴确乎乃小国寡民者，盖因其即无足够土地充疆扩土，只能令其子民居于层屋之上。”吾国人立论当与之不同，自具其理也。

而我必须向您坦言，亦无需假装要在两者间分出高下，但因中国营建宫室之法的确令我着迷。自我来华居住以来，我的审美及意趣也有了些许“中国味”呢！朋友之间说实在的，杜伊勒里宫[1]对面波旁公爵夫人府

① 杜伊勒里宫（Tuilleries）为法国皇家园林，位于巴黎塞纳河右岸。它始建于1564年，历史上屡经扩建，是在巴黎居住的大多数法国君主的行宫（从亨利四世至拿破仑三世时期），直至1871年巴黎公社时被焚毁。

" Emperor *Cang-hy*, whilſt he was
" looking over ſome Plans of our
" *European* Houſes,) this *Europe*
" muſt be a very ſmall and pitiful
" Country; ſince the Inhabitants
" cannot find Ground enough to
" ſpread out their Towns, but are
" obliged to live up thus in the
" Air." As for us, we think otherwiſe; and have Reaſon to do ſo.

However I muſt own to you,
without pretending to decide which
of the two ought to have the Pre-
ference, that the Manner of Building in this Country pleaſes me very
much. Since my Reſidence in
China, my Eyes and Taſte are grown
a little *Chineſe*. And, between
Friends, is not the *Ducheſs* of *Bour-
bon*'s Houſe oppoſite to the *Tuilleries*,
extreme-

邸[1]是不是相当漂亮？而且它仅一层，还在很多方面都合乎中国建筑的形制。每个国家皆有其审美趣味及习俗。吾国建筑之美无可争议，因其宏伟高大无出其右者。必须承认，吾国建筑之布局规划得非常合理。吾国建筑之所有部分皆遵循整体性与对称性，其间无突兀者，各部之间皆有呼应，相得益彰，和谐融洽，统一协调。美的秩序及布局，这在中国建筑中亦有之，特别如我在此信开头所述及的，帝都北京城的宫殿，更是如此。帝王宫殿、王公府第、官衙廨舍、乃至

① 此处波旁公爵夫人的府邸指波旁宫。

extremely pretty? Yet that is only of one Story, and a good deal in the *Chinese* Manner. Every Country has it's Tafte and Cuftoms. The Beauty of our Architecture cannot be difputed: nothing is more Grand and Majeftic. I own too, that our Houfes are well difpos'd. We follow the Rules of Uniformity, and Symmetry, in all the Parts of them. There is nothing in them unmatch'd, or difplaced: every Part anfwers it's Oppofite; and there's an exact Agreement in the Whole. But then there is this Symmetry, this beautiful Order and Difpofition, too in *China*; and particularly, in the Emperor's Palace at *Pekin*, that I was fpeaking of in the Beginning of this Letter. The Palaces of the Princes and great Men, the Courts of Juftice, and the Houfes

民间富户，屋宇皆遵循了中轴对称、方正严整的规制。

但他们的园林则一改常律*，从艺术条条框框的桎梏中解脱开来，追求不规则的美感。他们完全遵循这一原则，“他们在这里所要表现的，是天然野趣、幽隐逸趣，而不是严整之宫殿所遵循的艺术原则。”我还尚未在这

* 此信作者似乎已从他所处的中华帝国皇家园林中形成了自己的观点，但并非所有帝王的行宫园林都是如此。我最近亦见到一些他描绘的另一个花园的图像材料（寄自中华帝国，并很快会在此出版刊行），其中的地面、海子和园林，的确布置得很不规则，但屋宇、桥梁和围栏，却都是规则的。这些版画会给我们以最接近真实的理解，我们能就此体验中国的林泉之乐。

Houſes of the better ſort of People, are generally in the ſame Taſte.

But in their Pleaſure-houſes, they rather chuſe * a beautiful Diſorder, and a wandering as far as poſſible from all the Rules of Art. They go entirely on this Principle, " That
" what they are to repreſent there,
" is a natural and wild View of the
" Country; a rural Retirement, and
" not a Palace form'd according to all

* The Author of this Letter ſeems here to have form'd his Opinion, only from the Garden in which he was employ'd; for this is not univerſally the Caſe in the Pleaſure-houſes of the Emperor of *China*. I have lately ſeen ſome Prints of another of his Gardens, (brought from that Kingdom, and which will very ſoon be publiſh'd here,) in which the Diſpoſition of the Ground, Water, and Plantations, is indeed quite irregular; but the Houſes, Bridges, and Fences, are all of a regular Kind. Thoſe Prints will give the trueſt Idea, we can have, of the *Chineſe* Manner of laying out Pleaſure-grounds.

" the

壮阔的园林之中发现有任何两座小宫室是互相雷同的，无论它们相隔多远。或许您会认为，他们是从别的国家习得了各种样式，或他们的营造都过于随意，所造园林之各部分并无甚关联。当您读至此处时，可能会倾向于想象这样的作品非常可笑，甚至有碍观瞻。殊不知当您身临其境地注视它们，就会折服于这些不规则的艺术。所有绝佳的意境，都被精心营造，其美会渐次呈现。

" the Rules of Art." Agreeably to which, I have not yet obſerv'd any Two of the little Palaces in all the grand Incloſure, which are alike, tho' ſome of them are placed at ſuch conſiderable Diſtances from one another. You would think, that they were form'd upon the Ideas of ſo many different foreign Countries; or that they were all built at random, and made up of Parts not meant for one another. When you read this, you will be apt to imagine ſuch Works very ridiculous; and that they muſt have a very bad Effect on the Eye: but was you to ſee them, you would find it quite otherwiſe; and would admire the Art, with which all this Irregularity is conducted. All is in good Taſte; and ſo managed, that it's Beauties appear gradually, one after

见微知著，每一处景致也都值得细细玩味，您会久久徘徊其间，去发现它们的匠心独运，并满足您所有的好奇心。

另外，宫殿本身（纵然我已经提示您，相对整座园林而言，他们体量甚小）却也绝非微不足道之物。我曾于上一年亲见某位皇子为在这园林中增筑一栋建筑，费资近二十万英镑*，这还没有包括任何屋内的装修和陈设，因为这些都是由皇帝赏赐的。

* 原文为 Soixante Ouanes：在备注中，一乌瓦纳值一万两银子；每两银子值七个半里弗；所以那六十乌瓦纳值四百五十万里弗，相当于十九万六千八百七十五英镑。

after another. To enjoy them as one ought, you ſhould view every Piece by itſelf; and you would find enough to amuſe you for a long while, and to ſatisfy all your Curioſity.

Beſide, the Palaces themſelves (tho' I have called them little, in Compariſon of the Whole,) are very far from being inconſiderable Things. I ſaw them building one in the ſame Incloſure, laſt Year, for one of the Princes of the Blood; which coſt him near * Two hundred thouſand Pounds: without reckoning any

* The Original ſays, *Soixante Ouanes:* and adds in a Note, that one *Ouane* is worth Ten thouſand *Taëls*; and each *Taël* is worth Seven Livres and a Half; ſo that Sixty Ouanes make Four Millions and a Half of Livres. Which is equal to 196,875 Pounds Sterling.

thing

我必须就园林胜景的丰富性再赘述一句。其精彩并不止于格局、景致、规划、大小、高低以及总体方面的匠心。在那些细节之处，也同样重视此道。由是观之，除了在中华帝国，当世再无可能做出如此丰富的形式。仅门窗而言，它们就有圆形、椭圆形、方形，以及各种各样的多边形，一些犹如扇面，而另一些则是花、瓶、鸟、

thing for the Furniture and Ornaments of the Inside; for they were a Present to him from the Emperor.

I muſt add one Word more, in relation to the Variety which reigns in theſe Pleaſure-houſes. It is not only to be found in their Situations, Views, Diſpoſition, Sizes, Heights, and all the other general Points; but alſo in their leſſer Parts, that go to the compoſing of them. Thus, for inſtance, there is no People in the World who can ſhew ſuch a Variety of Shapes and Forms, in their Doors and Windows, as the *Chineſe*. They have ſome round, oval, ſquare, and in all Sorts of angled Figures; ſome, in the Shape of Fans; others in thoſe of Flowers, Vaſes, Birds, Beaſts, and

G Fiſhes;

兽、鱼的形状；简而言之，无论规则或不规则，形状几乎包罗万象。

也只有在此处，我相信，才可见到我将向你描述的这种廊庑。它们将宫殿各部分的房屋联系起来，在建筑之间延展开去。这些廊庑有时一边朝向屋舍，另一边在墙上开出各种花样的漏窗；也有仅列柱于廊两侧者。从所有宫殿皆可经由这些廊庑通向亭子，以纳清凉。这些游廊独一无二的妙处在于极少采用直线，而是百转

Fiſhes; in ſhort, of all Forms, whether regular or irregular.

It is only here too, I believe, that one can ſee ſuch Portico's, as I am going to deſcribe to you. They ſerve to join ſuch Parts of the Buildings in the ſame Palace, as lie pretty wide from one another. Theſe are ſometimes raiſed on Columns only, on the Side toward the Houſe; and have Openings, of different Shapes, thorough the Walls on the other Side: and ſometimes have only Columns on both Sides; as in all ſuch as lead from any of the Palaces, to their open Pavilions for taking the freſh Air. But what is ſo ſingular in theſe Portico's or Colonnades is, that they ſeldom run on in ſtrait Lines; but make an hundred Turns and

2 Wind-

迂回，时而穿林木而过，时而周行池岸。没有什么能如这些廊庑令人愉悦的了，它们营造出的田园氛围引人入胜，令人心旷神怡。

行文至此，您或许能从中了然我所陈述之事物。这个花费巨资营造而成的园林，除君临天下的中华帝国皇帝之外，再无人可称此财力，并且能在如此短的时间内营建完成。整座园林仅仅二十年就已颇具规模，由现任皇上的父亲肇造，现在，他的儿子，亦即当今圣上，

Windings: ſometimes by the Side of a Grove, at others behind a Rock, and at others again along the Banks of their Rivers or Lakes. Nothing can be conceiv'd more delightful: they have ſuch a rural Air, as is quite raviſhing and inchanting.

You will certainly conclude from all I have told you, that this Pleaſure-place muſt have coſt immenſe Sums of Money; and indeed there is no Prince, but ſuch an one as is Maſter of ſo vaſt a State as the Emperor of *China* is, who could either afford ſo prodigious an Expence, or accompliſh ſuch a Number of great Works in ſo little time: for all this was done in the Compaſs of Twenty Years. It was the Father of the preſent Emperor who began it; and his Son now

G 2 only

仅需在园内加些便利设施和装饰物件而已。

没什么更令人讶异的了。尽管所有殿宇皆为一层，他们雇佣了大量的宫人，就让工程以不可思议的速度推进起来。当物料到达施工现场之前，一半以上有难度的工艺已经完成了。在现场，工匠们将物料迅速有序地搭建起来，仅几个月，工程就竣工了。这种工程奇观，就如同仙宫楼阁忽现于奇山异谷之间或群峰之巅。

only adds Conveniences and Ornaments to it, here and there.

But there is nothing ſo ſurpriſing, or incredible, in this: for beſides that the Buildings are moſt commonly but of one Story, they employ ſuch prodigious Numbers of Workmen, that every thing is carried on very faſt. Above half the Difficulty is over, when they have got their Materials upon the Spot. They fall immediately to diſpoſing them in Order; and in a few Months the Work is finiſh'd. They look almoſt like thoſe abulous Palaces, which are ſaid to be raiſed by Inchantment, all at once, in ſome beautiful Valley, or on the Brow of ſome Hill.

This

整座宫苑，名曰“圆明园”，意为“园中之园”，或“万园之园”。属于君主的园林并非仅此一处，而是还有三处，它们都属于同一类型，但都不及圆明园规模宏阔，精致壮美。另一座园林——长春园，是由当今皇帝的祖父康熙皇帝*肇造，现今皇太后及其臣属居住于此。其他王公皇子之园规制较小；而帝王之园，规模宏大。

* 康熙皇帝是在1660年开始他的统治的；其子雍正，即位于1722年；其孙乾隆，即位于1735年。

This whole Inclofure is called, *Yven-ming Yven*, The Garden of Gardens; or The Garden, by way of Eminence. It is not the only one that belongs to the Emperor; he has Three others, of the fame Kind: but none of them fo large, or fo beautiful, as this. In one of thefe lives the Emprefs his Mother, and all her Court. It was built by the prefent Emperor's Grandfather *, *Cang-hy*; and is called *Tchamg tchun yven*, or The Garden of perpetual Spring. The Pleafure-places of the Princes and Grandees are in Little, what thofe of the Emperor are in Great.

* *Cang-hy* began his Reign in 1660; his Son, *Yongtching*, fucceeded him in 1722; and his Grandfon, *Kien-long*, in 1735.

Perhaps

或许您会问我，“为什么要花费如此冗长的笔墨去描述呢？何不绘制出这个宏伟园林的规划图，一并寄给您呢？”如若这样做，即便我终年无事，专事此图，也至少要花上三年的时日。实际上，我已经一刻不得闲暇，就连给您写信的时间，也是牺牲了休憩挤出来的。另外，如要完成这件画作，还需获准可自由出入园林，这非常不切实际。好在我略通绘事，否则，我就会与

Perhaps you will aſk me, "Why "all this long Deſcription? Should "not I rather have drawn Plans of "this magnificent Place, and ſent "them to you?" To have done that, would have taken me up at leaſt Three Years; without touching upon any thing elſe: whereas I have not a Moment to ſpare; and am forced to borrow the Time in which I now write to you, from my Hours of Reſt. To which you may add, that for ſuch a Work, it would be neceſſary for me to have full Liberty of going into any Part of the Gardens whenever I pleas'd, and to ſtay there as long as I pleas'd: which is quite impracticable here. 'Tis very fortunate for me, that I had got the little Knowlege of Painting that I have:

之前来华的几位耶稣会士那般，虽曾在这里居住二三十年，却未能涉足这宏阔园林一步，一窥堂奥。在这儿只有一个人，那就是皇帝，所有欢愉都是为了取悦他。这些迷人之地，除了皇帝本人及其后宫的妃嫔宫女太监等，其他人罕能得以一见。西人来华，唯钟表匠及画师可得入内，

have: for without this, I ſhould have been in the ſame Caſe with ſeveral other *Europeans*, who have been here between Twenty and Thirty Years, without being able ever to ſet their Feet on any Spot of this delightful Ground.

There is but one Man here; and that is the Emperor. All Pleaſures are made for him alone. This charming Place is ſcarce ever ſeen by any body but himſelf, his Women, and his Eunuchs. The Princes, and other chief Men of the Country, are rarely admitted any farther than the Audience-Chambers. Of all the *Europeans* that are here, none ever enter'd this Incloſure, except the Clock-makers and Painters; whoſe Employments make it neceſſary that they

探览各处，皆因职业使然。我工作的画室是我曾提及的一个小小的宫室，皇帝几乎每日来此巡视，这样一来，我们皆不敢懈怠。一般情况下，我们不得逾越该宫室之界一步，除非他们无法将需要我们绘画的题材带到我们面前，在这种情况下，他们会安排大批太监卫士引领我们去到画画的地方。我们须急步前驱，皆感如履薄冰，不能发出些微声息。我因此得入各处庭院，得观园林

they ſhould be admitted every where. The Place uſually aſſign'd us to paint in, is in one of thoſe little Palaces above-mentioned; where the Emperor comes to ſee us work, almoſt every Day: ſo that we can never be abſent. We don't go out of the Bounds of this Palace, unleſs what we are to paint cannot be brought to us; and in ſuch Caſes, they conduct us to the Place under a large Guard of Eunuchs. We are obliged to go quick, and without any Noiſe; and huddle and ſteal along ſoftly, as if we were going upon ſome Piece of Miſchief. 'Tis in this Manner that I have gone through, and ſeen, all this beautiful Garden; and enter'd into all the Apartments. The Emperor uſually reſides here Ten Months in each Year.

各处盛景。御驾驻跸是园，每年概有十月之久。我们距离北京城大约十英里，终日徜徉园林之内。圣上专置条案让我等从事绘事，晚间退出，住在自己置办的房屋中，近圆明园的入口。帝王返京，我们则随侍其左右，白天入大内绘画，夜归法国教堂住宿。*

此信行文至此，是时候搁笔了。于我而言，还是第一次写如此长的信札。希望它能给阁下带来愉悦，我也

* 这里原本有 14 至 15 页内容，因涉及作者的私人事务，或特派团的事务，与皇家园林无关，翻译因此省去。

Year. We are about Ten Miles from *Pekin*. All the Day, we are in the Gardens; and have a Table furniſhed for us by the Emperor: for the Nights, we have bought us a Houſe, near the Entrance to the Gardens. When the Emperor returns to *Pekin*, we attend him; are lodg'd there within his Palace; and go every Evening to the *French* Church *.

I think it is high time, both for you and me, that I ſhould put an End to this Letter; which has carried me on to a greater Length, than I at firſt intended. I wiſh it may give you any Pleaſure; and ſhould

* Here follow 14 or 15 Pages in the Original which treat only of the Author's private Affairs, or of the Affairs of the Miſſion, without any thing relating to the Emperor's Garden; and are therefore omitted by the Tranſlator.

H be

乐于竭尽所能，向您表达我对您的敬重。我将永远为您祷告，恳求您能够时常记起我。

鄙人，
带着最崇高的敬意，
向阁下致礼，

你最听话的，
卑微的仆人，

王致诚

be very glad if it was in my Power to do any thing more confiderable, to fhew you the perfect Efteem I have for you. I fhall always remember you, in my Prayers; and beg you would fometimes remember me in yours.

I am,

With the greateft Regard,

SIR,

Your moft obedient,

Humble Servant,

ATTIRET.

王致诚生平年表

1702年7月31日	1岁	让·德尼·阿蒂雷(Jean Denis Attiret)生于多尔，为一画家之子。
1735年7月31日	33岁	作为助理修士进入阿维尼翁耶稣会士初修院，期间为阿维尼翁大教堂做绘画工作。
1738年1月8日 1738年8月8日	36岁	乘船启程前往并抵达中国，取中文名为王致诚。
1739年/乾隆四年	37岁	七月二十七日，王致诚奉命为“万方安和”挂屏画油画。 九月初三日，王致诚与郎世宁画画，王幼学为“致松风开”画油画书格二张。 九月二十六日，王致诚奉命与画人张维邦在启祥宫行走。
1740年/乾隆五年	38岁	正月十一日，郎世宁、王幼学、王致诚等因画“元日舒长”画受赏。
1741年/乾隆六年	39岁	七月二十四日，郎世宁师徒五人奉命往瀛台、澄怀堂、长春书屋画油画。
1742年/乾隆七年	40岁	三月十二日，王致诚奉命照画稿一件画玻璃油画。 四月十四日，王致诚奉命进如意馆画玻璃画。 四月二十二日，又奉命为“方壶胜境”画玻璃油画。 五月十七日，王致诚奉命画油画玻璃斗方八块。

七月初一日，太监高玉交钟表套一件，交王致诚认看、找画。
七月初十日，王致诚奉命画玻璃画一块。
八月二十三日，王致诚获赏缎一匹。时郎世宁获赏元宝一个；王幼学、张为邦、丁观鹏、沈源均获赏缎一匹。
九月初四日，郎世宁、王致诚为建福宫小三卷房床罩内玻璃镜画油画花卉起稿。奉旨："照样准画，其玻璃着郎世宁在本处画。"
十月初二日，王致诚奉命在造办处油画房画建福宫小三卷房床罩玻璃油画。
十月十二日，郎世宁等为"汇芳书院"画油画。
十月二十六日，太监胡世杰交郎世宁画十骏绢画十张，着托裱大画十轴。此《十骏图》裱成之后，深受乾隆之喜爱，命交懋勤殿，让翰林评定等级，并特制黑红漆画金龙箱收贮。按《十骏图》共十轴，俱长一丈六尺六寸、宽九尺八寸五分，由张照等按月分编次，俱评为上等。计：月一，万吉骦；月二，阚虎骝；月三，狮子玉；月四，霹雳骧；月五，雪点凋；月六，自在骕；月七，奔霄骢；月八，赤花鹰；月九，英骥子；月十，尔云驶。至乾隆十三年四月，又将郎世宁绘《大宛骝图》《红玉座图》和《如意骢图》三轴续入《十骏图》内。

1743年／乾隆八年　41岁　三月十二日，命王致诚照画稿将一无架紫檀木边玻璃小插屏镜上锡片刮下来画油画。

五月七日，太监胡世杰持来玻璃斗方八块，命王致诚画油画。

是年阳历11月1日，王致诚致函欧洲，详述与郎世宁在如意馆之绘画生活："吾人所居乃一平房，冬寒夏热。视为属民，皇上恩遇之隆，过于其他传教士。但终日供奉内廷，无异囚禁，主日瞻礼，亦几无祈祷暇晷。作画时颇受掣肘，不能随意发挥。"

1744年／乾隆九年　42岁　二月初十日，王致诚画长春仙馆中层明间面阔板墙中间油画表盘一张。

六月初四日，命王致诚画油画。

1745年／乾隆十年　43岁　二月初一日，太监高玉等交钟表套一件，着交王致诚看画得画不得。

六月初五日，命王致诚画油画。

1746年／乾隆十一年　44岁　三月初一日，命画油画人王幼学等往香山行宫贴画。十四日，太监胡世杰持来《扫象图》一幅，传旨丁观鹏用宣纸画一幅。二十三日，命王致诚为紫檀木边玻璃挂屏画花卉。

十二月，命王致诚为建福宫屏风后二扇上画美人图二张。

1747年/乾隆十二年　45岁

五月初十日，王致诚为香山"情赏为美"圆光门上画美人图一幅。

十一月初一日，命照噶尔丹策楞进的马，着郎世宁、王致诚、艾启蒙各画马一批。

是年，在巴黎出版了王致诚的《帝都来信：北京皇家园林概览》，描绘圆明园甚详，为庭园爱好者所争诵，给予英国庭园界以相当刺激。

1748年/乾隆十三年　46岁

五月二十六日，命王致诚画美人图；命郎世宁为养心殿后三殿三面墙棚顶起通景画稿。

闰七月初一日，命王致诚画簾刷下画美人面向右。

1749年/乾隆十四年　47岁

六月二十二日，王致诚画得同乐园双美人画稿一张，至七月十二日竣。

七月十五日，王致诚再画圆光美人画稿一幅。

九月，郎世宁、王致诚、艾启蒙、丁观鹏、周鲲、张为邦、姚文翰、余稚、王幼学、张廷彦、张廉、福荣、伊兰太、德舒、福贵、福海等奉命画斗方，共计七十九幅。

十一月初九日，王幼学画瀛台兰室通景画，王致诚画背面通景稿。王致诚又奉命画六方亭。丁观鹏、余省、周鲲、张镐画《雪景守岁图》。冷枚画《汉宫春晓》。

十二月十四日、十九日、二十七日，三次传旨安排瀛台兰室通景画稿，王幼学奉命照瀛台兰室通景画稿画画，背面由王致诚起通景小稿，奉旨准画。于十二月二十九日，王致诚完成西边门上小样、东边板墙上小样各一张。

1749 年 / 乾隆十四年　47 岁

《帝都来信：北京皇家园林概览》英文版出版。

1750 年 / 乾隆十五年　48 岁

正月，王致诚画“芙蓉坪”东壁横披一张。

十月初六日，王致诚画西洋式画屏香几花样。周鲲、张镐画《热河全图》大画一幅。郎世宁原起的《热河总图稿》，着周鲲、张镐另起稿，俟准时再画绢画。郎世宁画长春园水法房正殿花盖、靠被、坐褥，发往南边成做。

十一月二十二日，王致诚、王幼

学画静宜园"云楼松坞"云庄殿通景画。

十二月初十日奉旨："民人私典旗地，定例綦严，屡经饬禁。但念郎世宁等系西洋远人，内地禁例原未经通饬遵行，且伊等寄寓京师，亦藉此以资生计，所有定例后价典旗地，著加恩免其撤回。入原典之人自行用价收赎，仍听其赎回。此系朕加恩远人，恩施格外。今禁例即经申明，嗣后西洋人于此项地亩之外，再有私行典买旗地者，与受之人定行照例治罪，并将此次恩免撤回之处从重究治，郎世宁等既经宽免，所有典出之蔡永福等，并失察之该管各官，均从宽免其治罪议处。至河淤地亩，亦系郎世宁等加典之地，俱免圈撤。但蔡永福于认买公产之外，所有多得河淤地亩典价，并非伊分内应得之项，着该部照例查办，钦此。"①

1751年/乾隆十六年　49岁

六月二十九日，郎世宁画得西洋

① 按此项上谕，被在京西洋人立石刻碑，俗称"价典旗地许可碑"，1911年9月发现于北京长辛店附近天主堂旧址，文仅109字。

法字陈设纸样，交西洋人杨自新（原名 Ir. Gilles Thebault，1738年入华，1766年卒，法国人）、席澄源带领役匠在如意馆制作。

又命郎世宁仿西洋铜版画手卷款式，为长春园水法房大点三间、东西梢间四间、游廊十八间、东西亭子二间、顶棚连墙，起通景画稿。至当年十一月画稿完成，又命王致诚照稿放大。

十二月初七日，郎世宁将“万年欢”陈设更改做法，烫得合牌小样一件，奉旨：“水法座子改铜胎撒镀金，鸳鸯做珐琅的。”[①]

1752年/乾隆十七年　50岁

二月二十一日，王致诚奉命为油画菊花玻璃找补梗叶。

四月初二日，王致诚为“九州清晏”画美人图一幅。

1753年/乾隆十八年　51岁

十一月初八日，郎世宁奉命照西洋铜版画手卷二卷为长春园水法房大点三间、东西间四间、游廊十八间、东西亭子二间，俱起通

① 按此件“万年欢”是专为庆贺皇太后七十寿辰而制作的机械人表演的戏剧模型。此模型献上之后甚得太后与乾隆的喜欢，乾隆帝曾亲自到如意馆仔细观看，提出具体的修改要求，并对参与制作者均给予厚赏，因而使在京供职的西洋耶稣会士们大得体面。

景画稿四张。奉旨："着王致诚放大稿。"

1754年 / 乾隆十九年　52岁

正月二十四日，郎世宁查得现有画过脸像西洋人十四个，呈览。奉旨："着是死人之内拣去三个，将现在内庭行走郎世宁等六十（人）画上，钦此。"①

三月初七日，郎世宁奉命将张为邦从热河临来焦秉贞稿放大，起通景画稿，众徒弟们帮画，着王致诚画脸像。

四月十六日，王致诚为万寿山乐安和圆光门二面画美人。

五月初七日，王致诚奉命往热河画油画十二幅。

六月初八日，命王致诚画油画御容一幅，放热河悬挂。

七月二十三日，命郎世宁、王致诚、艾启蒙等前往热河，派库掌六达子并如意馆柏春阿一名，将郎世宁等应用骡脚钱粮，妥协办理送往。

九月二十三日，与郎世宁、艾启蒙前往热河。

十月初三日，命王致诚画油画御

① 按此举表明乾隆帝对传播西洋文明的西洋人的重视和礼遇。

容一幅，热河挂。
十月二十八日，王致诚来热河画油画脸像十幅。

1755年/乾隆二十年　53岁

五月初九日，郎世宁、王致诚、艾启蒙奉命画筵宴图大画二幅，于七月初十日松热河张贴，贴在“卷阿胜境”内东西两山墙上。
五月十九日，将王致诚画的御容射箭吊屏一幅松热河悬挂。
八月十四日，郎世宁、王致诚、艾启蒙三人奉命去热河画油画，并令带上油画脸像。画得脸像六份存如意馆内。

1756年/乾隆二十一年　54岁

四月初六日，传旨着席澄源照王致诚画的跑马挂屏上圈筒做一个小样，俟候呈览。
四月初七日，太监胡世杰传旨：“长春园‘谐奇趣’东边，着郎世宁起西洋式花园地盘样稿呈览，准时交圆明园工程处成造，钦此。”于本日郎世宁起得西洋式花园小稿一张，呈览，奉旨：“照样准造，其应用西洋画处，着如意馆画通景画，钦此。”
六月初四日，命王幼学为“乐安和”向西门画线法画，着王致诚添画人物脸像。又命郎世宁照库理狗的坐像画画一张。

七月十三日，张廷彦奉命在长春园全图上添画蒨园、“谐奇趣”及新添西洋水法，有不明白处问郎世宁画。此《长春园全图》至乾隆二十二年五月二十五日告竣。

八月十六日，传旨着郎世宁、王致诚、丁观鹏、姚文翰，仿刘宗道画《照盆孩儿》各画一张。于闰九月十五日画完呈进。

1757年/乾隆二十二年　55岁

十一月初一日，命王致诚为玻璃灯画画，并要求“往热闹里添画”。

十二月十四日，命张为邦、王致诚各画牛鹿图一张。

1758年/乾隆二十三年　56岁

正月二十一日，命王致诚照静宜园白驼鹿画一幅，仍用旧胎骨换裱。

六月初五日，命王致诚为澄虚榭仿郎世宁夜景横披画绢画一幅。

六月二十二日，王致诚奉命为含经堂影壁画油画狮子、老虎。

十月十五日，王致诚为爱山楼后泽兰堂画油画美人一张，先起稿呈览。

十一月初十日，王致诚画线法画一张。

1759年/乾隆二十四年　57岁

二月十二日，王致诚画得夜景大画并油画美人，俱着托贴。

三月二十日，命王致诚为无福堂

画遭风船景致绢画一张。

三月二十六日，王致诚为长春园澄观阁画门及盘山画门绢画美人三张。

四月十九日，王致诚新画遭风船画一张。

十二月十五日，太监胡世杰传旨命王致诚为澄观堂后殿东间画美人图。

十二月十六日，太监胡世杰传旨命郎世宁、艾启蒙、王致诚画众大人油画脸像。

1760年/乾隆二十五年　58岁

三月二十五日，郎世宁为新建水法西洋门内八方亭画西洋画；王致诚为新建水法三间楼上画绢画任务四幅。

四月十一日，王致诚为承光殿古籁堂画油画假门二张。

五月初九日，命新来西洋人①内有会水法的，着收拾交泰殿铜壶滴漏。

六月二十三日，王致诚奉命为“蓬岛瑶台”宝座格后面油画门三扇画西洋水法风景油画，并为紫檀木插屏一座另画油画一张。

十月十三日，王致诚奉命为热河“夕佳楼”画西洋通景画大画。

① 按此西洋人当为韩国英。

- 1761年/乾隆二十六年　59岁　四月初九日，命郎世宁为新建水法十一间楼下北明间二间、南明间二间，四面俱用苏州织来白毯子，照原来西洋毯子仿画。
五月十三日，命郎世宁为新建水法十一间楼明间内西进间周围墙连棚顶，起稿画通景画。
五月十六日，命王致诚为"谐奇趣"楼上西平台九屏风背后画堂画西洋油画、水画厄鲁特回子稿，交柏堂阿等画。
- 1762年/乾隆二十七年　60岁　闰五月十五日，太监胡世杰命王致诚为翔凤艇五舱左夹道北墙门上画美人图。
六月初三日，王致诚为思永斋画西洋人物绢画一幅。
十月二十六日，命王致诚为怡性轩楼下用绢画人物，王幼学补景。
十二月初十日，西洋人钱德明[1]认看表一件、千里眼一件。
- 1763年/乾隆二十八年　61岁　十二月二十五日，命王致诚为思永斋西楼梯上拉门一道东面画美人画，西面着王幼学画书格一张。
正月二十三日，郎世宁为思永斋

[1] 钱德明，字若瑟，原名Joan-Joseph Maria Arniiot，法国人，1751年入京，1793年卒于北京，著有《满蒙文法 满法字典》《汉满蒙藏法五国文字字汇》《中国历代帝王纪年表》《孔子传》等书。

东暖阁镶嵌罩配画格起稿，着王幼学用绢画；王致诚为瀛台听鸿楼画西洋画一幅。

三月二十八日，艾启蒙奉命修补“谐奇趣”大殿棚顶画；王致诚、王幼学为思永斋画通景画绢画；郎世宁奉命画关防绢画。

五月初九日，接秀山房澄练楼楼下南间南墙着郎世宁画御容，方琮补景，绢画一幅。南间东墙，王致诚、王幼学等画线法绢画一幅。北间北墙着姚文瀚、陆遵书画九峰园图绢画一幅。

1764年/乾隆二十九年　62岁

正月二十三日，王致诚等为水法十一间楼画挂屏、门斗、窗户斗等。

四月二十二日，交郎世宁《群马图》挂轴一轴，配囊。王致诚奉命为瀛台日知阁殿内五屏风群板画宣纸异兽。

五月二十九日，王致诚、王幼学为“澄虚榭”静香馆画支窗线法绢画一幅。王致诚、王幼学为“蓬岛瑶台”两卷房西里间南墙起画稿，人物着金廷标画，线法景着王致诚画绢画一幅。

十月十七日，王致诚为“玉玲珑

馆"后殿西墙线法画面周围添画。

1765年/乾隆三十年　63岁

五月二十六日，奉旨："平定准噶尔、回部等处得胜图十六幅，着郎世宁等绘画底稿，发往西洋，拣选能艺，依稿刻做极细铜版。其铜版不可靬做，所用工料任其开报，如数发给。令将郎世宁画得《爱玉史诈营稿》一张、王致诚画得《阿尔楚尔稿》一张、艾启蒙画得《伊利人民投降图》一张、安得义画得《库尔满稿》一张先行发去，作速刻作，得时每版用整纸先刷印一百张，随铜版一同交来。其余十二章陆续三次发去，钦此"。

六月十六日，将郎世宁等四人所画得胜图稿十六张内，先画得四张并汉字旨意帖一件、西洋字帖四件交太监胡世杰呈览，奉旨："着交王贵常交军机处发往粤海关监督方体浴遵照办理，钦此。"

七月初七日，太监胡世杰传旨："艾启蒙起的铜版稿子，得时交金廷标画；王致诚起的稿子，得时著姚文翰画。"

1766年/乾隆三十一年　64岁

五月十五日，赏郎世宁菜一桌，王致诚、艾启蒙、安得义菜一桌

年份	年龄	事件
		半，共二桌半。每桌素菜两碗、摊鸡蛋一碗、虾米白菜一碗、又点心一盘及素粉汤。（按这是目前查到的有关郎氏最后一次在清宫如意馆活动的记录。）
1766年／乾隆三十一年	64岁	六月初十日，郎世宁病逝（78岁）。
1767年／乾隆三十二年	65岁	十月初五日，太监胡世杰传旨命王致诚画盘山引胜西洋水法门口美人图，含远楼门口旧画接补。
1768年／乾隆三十三年	66岁	二月十九日，王致诚画水法三间楼上插屏人物绢画二幅。 三月二十八日，王致诚画鑑园师善堂北次间北墙西边假门上人物绢画一幅。
1768年12月8日／乾隆三十三年	66岁	王致诚在北京去世。

注：本年表主要依据《清中前期西洋天主教在华活动档案史料（1–4卷）》《清宫内务府造办处档案总汇》和《清宫廷画家郎世宁年谱 兼在华耶稣会士史事稽年》编绘。凡涉及中国纪年，以朝代年号纪年、农历（个别情况）为准；凡涉及其他年份，以公历为准。如论述传教士来华之前的，以公历为准；论述传教士来华之后的，以清朝年号纪年为准；论述涉及清宫活计档者，以农历为准。

清宫档案中的王致诚档案节选

乾隆四年九月初三日 九一

初三日，画画人戴正持来押帖一件，内开本日太监毛团传旨：着西洋人王致诚画画，柏唐阿、王幼学至松风阁画油画书格二张，钦此。

乾隆四年九月二十六日 九二

二十六日，催总廖保持来司库郎正培押帖一件，内开本日太监毛团传旨：西洋人王致诚、画画人张维邦等着在启祥宫行走，各自画油画几张，钦此。

乾隆七年八月二十三日 一一二

二十三日，司库白世秀来说，太监高玉等交银元宝一个、缎六疋，传旨：赏郎世宁元宝一个，其王致诚、王幼学、张为邦、孙祜、丁观鹏、沈源每人缎一疋。钦此。

乾隆八年三月十二日 一一八

十二日，司库郎正培交无架紫檀木边玻璃小插屏镜一件、随画稿一件，传旨：着西洋人王致诚照画稿将镜上锡片刮下画油画，钦此。

乾隆八年五月十七日 一二〇

十七日，副催总六十七持来司库郎正培、骑都尉巴尔党催总花善押贴一件，内开为本月初七日，太监胡世杰持来玻璃斗方八块，传旨：着王致诚画油画，钦此。

乾隆九年二月初十日 一二八

初十日，副催总六十持来司库郎正培、骑都尉巴尔党、催总花善押贴一件，内开为正月二十四日，太监胡士杰交长春仙馆墨池云装修纸样一张，传旨：中层明间面阔板墙中间，着王致诚速画油画表盘一张，钦此。

乾隆十年二月初一日 一四一

初一日，副催总王来学持来司库郎正培、骑都尉巴尔党押帖一件，内开八年七月初一日司库白世秀、首领萨木哈、副催总达子来说，太监高玉等交钟表套一件，传旨：着交与王致诚看画得画不得，钦此。

乾隆十年六月初五日 一四七

初五日，副催总王来学持来司库郎正培、骑都尉巴尔福押帖一件，内开为七月初一日太监胡世杰传旨：着王致诚画油画，钦此。

乾隆十二年五月初十日 一六二

初十日，副催总六十七持来司库郎正培、副司库瑞保押帖一件，内开为五月十一日太监胡世杰传旨：香山“情赏为美”圆广门上着王致诚画美人一幅，钦此。

乾隆十二年十一月十三日 一六七

十三日，副催总六十七持来司库郎正培、副司库瑞保押帖一件，内开为十一月初一日太监胡世杰传旨：将葛尔丹策楞进的马，着郎世宁、王致诚、艾启蒙各画马一批，钦此。

乾隆十三年五月二十六日 一七五

二十六日，司库白世秀、催总六达子来说，太监胡世杰传旨：启祥宫王致诚画簾刷下画美人面向右，钦此。

乾隆十三年闰七月初一日 一七八

初一日，副催总六十七持来司库郎正培瑞保押帖一件，内开为五月二十七日太监胡世杰交宁静舟门口样一张，传旨：着王致诚画簾刷下画美人面向右，钦此。

乾隆十四年七月十五日 一九四

十五日，副催总佛保持来员外郎郎正培、库掌瑞保押帖一件，内开为五月十二日太监王自云来说，太监张永泰传旨：同乐园圆光门内着王致诚画美人一幅，钦此。

于本月二十日，王致诚画得凄美人稿一张，员外郎郎正培、库掌瑞保呈览，奉旨：准照样一边画凄美人，一边是一字一画，钦此。

于八月十七日，员外郎郎正培奉旨：王致诚现画的圆光美人画留着别处用，再照尺寸画圆光美人一幅，钦此。

乾隆十四年十一月初九日 一九九

十一月初九日，副催总佛保持来员外郎郎正培、库掌瑞保押帖一件，内开为本月十七日造办处来帖，内开员外郎白世秀来说，太监胡世杰交瀛台兰室通景画稿一张，传旨：照纸样就着王幼学画，背面着王致诚起稿，钦此。

于本月十九日，王致诚画得背面通景小稿一张，郎正培、瑞保呈览，奉旨：照样准画，钦此。

于十月二十七日，太监胡世杰传旨：此通景画两边门关上前后要一式通景，开门两扇与中间亦要一式通景。背后两边桶子门里外与东边板墙俱画一式通景。着王致诚画样呈览，钦此。

于本日，王致诚起得六方亭稿纸样一张，太监卢成持去，交太监胡世杰呈览，奉旨：照样准画，前后糊顶不必画，钦此。

乾隆十四年十二月十四日 二〇〇

十四日，副催总佛保持来员外郎郎正培、库掌瑞保押帖一件，内开为十月十七日造办处来帖，内开员外郎白世秀来说，太监胡世杰交瀛台兰室通景画稿一张，传旨：照纸样就着王幼学画，画背面着王致诚起稿，钦此。

于本月十九日，王致诚画得背面通景小稿一张，郎正培瑞保呈览，奉旨：照样准画。钦此。

于本月二十七日，太监胡世杰传旨：此通景画两边门关上前后要一式通景，开门两扇与中间亦要一式通景，背后西边桶子门里外与东边板墙俱画一式通景，着王致诚画样呈览，钦此。

于本日，王致诚起得通景六方亭小稿样一张，太监卢成持去交太监胡士杰呈览，奉旨：照样准画，前面棚顶不必画。钦此。

于本月二十九日，王致诚画得西边门上小样一张、东边板墙上小样一张，郎正培、瑞保呈览，奉旨：照样准画，钦此。

乾隆十五年八月初一日 二〇四

初一日，副司库佛保持来员外郎郎正培、司库瑞保押帖一件，内开为十四年七月三十日，太监胡世杰交盘山行宫引胜轩小样一件，传旨：着西洋人照尺寸连棚顶画满四面通景样呈览，钦此。

于十月初六日，王志诚画得西洋式画屏香几花样，郎正培、瑞保呈览，奉旨：照样准画，钦此。

于十五年三月初六日，郎正培瑞保奉旨：此画上石青扁联朕写填金字，钦此。

乾隆十五年十月二十二日 二〇六

二十二日，副催总佛保持来员外郎郎正培、库掌瑞保押帖一件，内开为本月十七日太监胡世杰传旨：静宜园棲云楼松坞云庄殿内西进间南板墙上，着王致诚、王幼学起通景画稿呈览，钦此。

于本月二十日，王致诚、王幼学起得通景稿，太监卢成持去交太监胡世杰呈览，奉旨：照样准画，钦此。

乾隆十八年十一月初八日 二一四

初八日，副催总高五十持来员外郎郎正培、催总德魁押帖一件，内开十七年四月初一日为十五年六月初八日，太监胡世杰交西洋铜板手卷二捲，传旨：长春园水法房大殿三间、东西稍间四间、游廊十八间、东西亭子二间、棚顶连墙俱着郎世宁仿卷内款式起通景画稿呈览，钦此。

于十一月三十日，郎世宁起得通景画小稿四张，郎正培等呈览，奉旨：照样准画，着王致诚放大稿，钦此。

乾隆十九年四月初三日 二一八

初三日，副催总六十一持来员外郎郎正培、催总德魁押帖一件，内开为十九年三月初七日太监胡世杰传旨：瀴鹤斋前殿西墙将张为邦热河临来焦秉贞画稿着郎世宁、周圆放出照西墙尺寸起通景画稿，众徒弟们帮画，着王致诚画脸像，钦此。

乾隆十九年四月十六日 二二一

十六日，副催总六十一持来员外郎郎正培、催总德魁押帖一件，内开为三月初三日太监胡世杰传旨：万寿山乐安和圆光门二面着王致诚画美人图，钦此。

乾隆十九年七月二十三日 二三一

二十三日，副催总六十一持来员外郎郎正培、催总德魁押帖一件，内开为十九年五月初七日承恩公德奉旨：带领西洋人王致诚往热河画油画十二幅，钦此。

乾隆十九年九月二十三日 二三四

二十三日，奉怡亲王内大臣海谕令遵旨：着西洋人郎世宁、王致诚、艾启蒙前往热河，着派库掌六达子并如意馆柏唐阿一名将西洋人郎世宁等应用骡脚食物俱动用钱粮妥协办理送往，遵此。

乾隆十九年十月初三日 二三五

十月初三日，副领催六十一持来员外郎郎正培押帖一件，内开为本年六月初八日首领太监张玉传旨：着西洋人王致诚画油画御容一幅热河挂，钦此。

乾隆十九年十月二十八日 二三六

二十八日，员外郎郎正培奉旨：西洋人王致诚热河画来油画脸像十幅，着托高丽纸二层周围俱镶金边，钦此。

乾隆二十年四月二十四日 二四六

二十四日，副催总海陞持来如意馆押帖一件，内开二十年正月二十二日为乾隆十九年六月十三日，承恩公德保奉旨：着王致诚画御容射箭吊屏一幅，钦此。

于乾隆二十年正月二十一日，员外郎郎正培、催总德魁面奉上谕：将吊屏添宽三尺添高五寸，钦此。

于乾隆二十年五月十九日，员外郎郎正培将画得御容吊屏一件交内务府大臣三和请至热河交胡全忠挂讫。

乾隆二十一年八月十七日 二六四

十七日，接得员外郎郎正培、催总德魁押帖一件，内开本月十六日太监张永泰传旨：着郎世宁、王致诚、丁观鹏、姚文瀚仿刘宗道画《照盆孩儿》各画一张，钦此。

乾隆二十三年六月初五日 二八六

初五日，接得员外郎郎正培、催总德魁押帖一件，内开本月初四日太监胡世杰传旨：澄虚阁着王致诚仿郎世宁画过夜景横披，用白绢画一幅，钦此。

乾隆二十三年十月十七日 二九七

十七日，接得员外郎郎正培押帖一件，内开本月十五日太监胡世杰持来纸样一张，来说：爱山楼后泽兰堂大殿东次间隔断板中间安玻璃镜……北边向东新添假门口一座，净高五尺六寸三分宽二尺六寸八分，传旨：着王致诚画油画美人一张，先起稿呈览，钦此。

乾隆二十三年十一月初十日 二九九

初十日，接得员外郎郎正培押帖一件，内开本月初八日太监胡世杰交白宣纸一张，传旨：着王致诚按线画法画画一张，钦此。

乾隆二十四年二月十二日 三〇〇

十二日，接得员外郎郎正培、库掌德魁押帖一件，内开本日奉旨：王致诚画得夜景大画并油画美人俱着托贴，钦此。

乾隆二十四年三月二十日 三〇三

二十日，接得库掌德魁押帖一件，内开本月十九日太监胡世杰传旨：五福堂着王致诚照张照字横披尺寸画遭风船景致绢画一张，钦此。

乾隆二十四年三月二十六日 三〇五

二十六日，接得库掌德魁押帖一件，内开本月二十五日太监胡世杰传旨：长春园澄观阁画门二张、盘山画门一张，俱着王致诚用白绢画美人，钦此。

乾隆二十四年四月十九日 三〇六

十九日，接得员外郎郎正培、库掌德魁押帖一件，内开本月十八日奉旨：着金廷标画横披大画一张，得时在五福堂对张照字贴，其王致诚所画遭风船画得时另看地方贴，钦此。

乾隆二十四年十二月十六日 三一四

十六日，接得库掌德魁押帖一件，内开本月十五日太监胡世杰传旨：澄观堂后殿东间北边假门上换吴应枚画条一张，王致诚画美人，钦此。

乾隆二十四年十二月十七日 三一五

十七日，接得库掌德魁押帖一件，内开本月十六日太监胡世杰传旨：着郎世宁、艾启蒙、王致诚画众大人油画脸像，钦此。

乾隆二十五年三月二十五日 三一八

二十五日，接得员外郎安泰、金辉押帖一件，内开本月二十一日太监胡世杰传旨：新建水法三间楼上东西墙挂屏四面着王致诚用绢画人物四幅，钦此。

乾隆二十五年六月二十六日　　三二〇

二十六日，接得员外郎安泰、金辉押帖一件，内开本月二十三日太监胡世杰传旨：蓬岛瑶台宝座格后面油画门三扇着王致诚改画西洋水法景紫檀木插屏一座，另画油画一张，钦此。

乾隆二十五年十月二十二日　　三二四

二十二日，接得员外郎安泰、金辉押帖一件，内开本月十三日太监胡世杰传旨：热河夕佳楼殿内着王致诚画西洋式通景大画，钦此。

乾隆二十六年五月十六日　　三三六

十六日，接得员外郎安泰德魁押帖一件，内开本月十五日太监胡世杰传旨：谐奇趣楼上西平台九屏风背后画堂着画西洋油画、水画厄鲁特回子，俱着王致诚起稿，柏唐阿等画，钦此。

乾隆二十七年闰五月初一日　　三四五

初一日，接得郎中达子、员外郎安泰押帖一件，内开五月十六日首领董五经来说，太监胡世杰传旨：翔凤艇五舱左夹道北墙门上用画条一张，着王致诚画美人，钦此。

乾隆二十七年六月初三日　　三四九

初三日，接得员外郎安泰、库掌花善押帖，内开五月十九日太监胡世杰传旨：思永斋佳处领其要插屏一座，正面着金廷标仿宋人画宣纸画一幅，背面着王致诚用绢画西洋人物一幅，钦此。

乾隆二十七年十月二十六日　　三五二

十月二十六日，接得员外郎安泰、李文照押帖一件，内开本月十四日太监如意传旨：怡性轩楼下西一间揭下油画一张着王致诚画人物，王幼学补景，用绢画，钦此。

乾隆二十七年十二月二十五日 三五五

二十五日，接得员外郎安泰、李文照押帖一件，内开本月初九日太监如意传旨：思永斋西间楼梯上拉门一道，面东着王致诚画美人，面西着王幼学书格一张照九州清晏画格一样，钦此。

乾隆二十八年正月二十三日 三五七

二十三日，接得员外郎安泰、李文照押帖一件，内开本月二十一日太监如意传旨：瀛台听鸿楼北门内玻璃插屏后着王致诚画西洋画一幅，钦此。

乾隆二十八年三月十八日 三五九

十八日，接得员外郎安泰、李文照押帖一件，内开二月十八日太监如意传旨：思永斋东暖阁东西墙二面着王致诚、王幼学等合通景绢画，钦此。

乾隆二十八年五月初九日 三六三

初九日，接得员外郎安泰、李文照押帖一件，内开四月二十五日太监荣世泰传旨：接秀山房澄练楼楼下南间南墙着郎世宁画御容，方琮补景，绢画一幅，南间东墙着王致诚、王幼学等画线法，绢画一幅，北间北墙着姚文瀚、陆遵书画九峰园图，绢画一幅，钦此。

乾隆二十九年正月二十三日 三六九

二十三日，接得郎中德魁、员外郎安泰、李文照押帖一件，内开本月初八日起得水法十一间楼明三间殿内东西墙上挂屏四件、门斗窗户斗八件呈览，奉旨：着西洋人王致诚等画，钦此。

乾隆二十九年四月二十二日 三七三

二十二日，接得郎中德魁等押帖，内开本月十二日太监如意传旨：瀛台日知阁殿内五屏风峰群着王致诚画宣纸异兽，钦此。

乾隆二十九年五月二十九日 三七五

二十九日，接得郎中德魁等押帖，内开本月二十三日太监如意传旨：澄虚榭静香馆着王致诚、王幼学等画支窗线法绢画一幅，钦此。

乾隆二十九年五月二十九日 三七六

二十九日，接得郎中德魁等押帖，内开本月二十四日太监如意传旨：蓬岛瑶台新盖两卷房西里间南墙着王致诚、王幼学起稿呈览，钦此。

于二十四日起得人物线法稿一张呈览，奉旨：人物着金廷标画其线法，景着王致诚画绢画一幅，钦此。

乾隆二十九年十月十七日 三七七

十七日，接得郎中德魁、员外郎安泰、李文照押帖一件，内开本月十四日太监胡世杰传旨：玉玲珑馆殿西墙线法画上着王致诚等周围添画，钦此。

乾隆三十年七月十九日 三八七

十九日，接得郎中德魁、员外郎安泰押帖，内开七月初七日太监胡世杰传旨：艾启蒙起的铜板稿子得时即着金廷标画，王致诚起的稿子得时着姚文瀚画，钦此。

乾隆三十一年十月十一日 三九六

十一日，接得员外郎安泰押帖一件，内开十月初五日太监胡世杰传旨：盘山引胜轩西洋水法门口着王致诚画美人一张，含远楼门口旧画接补，钦此。

乾隆三十二年二月十九日 四〇〇

十九日，接得员外郎安泰、李文照押帖，内开本月十四日太监如意传旨：水法三间楼上插屏着王致诚画人物绢画二幅，钦此。

乾隆三十三年三月二十八日 四一一

二十八日，接得员外郎安泰押帖，内开本月十二日太监胡世杰传旨：鑑园师善堂北次间北墙西边假门上着王致诚画人物绢画一幅，钦此。

乾隆三十四年七月初二日 四二三

初二日，接得郎中李文照等押帖，内开六月二十九日太监胡世杰传旨：王致诚画过水法三间楼上插屏背面西洋人物画二幅，一幅画至八成，着艾启蒙接画，未画一幅不必画，钦此。

注：本节选选编自《清中前期西洋天主教在华活动档案史料（1-4卷）》中有关王致诚的条目。

段建强，同济大学建筑与城市规划学院建筑历史博士（2006—2012）、复旦大学文物与博物馆系高级访问学者（2009—2010）。现任教于河南工业大学。主要研究方向为中西方园林史、近代学科史、城市遗产保护及更新。曾参编《陈从周全集》（编委、主编第五卷）、译著《无限之境：法国十七世纪园林及其哲学渊源》、专著《从谐奇趣到明轩：十七至二十世纪中西文化交流拾遗》等。